Extrusion of Foods

Volume I

Author

Judson M. Harper
Professor
Agricultural and Chemical Engineering
Colorado State University
Fort Collins, Colorado

CRC Press
Taylor & Francis Group
Boca Raton London New York

CRC Press is an imprint of the
Taylor & Francis Group, an **informa** business

CRC Press
Taylor & Francis Group
6000 Broken Sound Parkway NW, Suite 300
Boca Raton, FL 33487-2742

Reissued 2019 by CRC Press

© 1981 by Taylor & Francis Group, LLC
CRC Press is an imprint of Taylor & Francis Group, an Informa business

No claim to original U.S. Government works

This book contains information obtained from authentic and highly regarded sources. Reasonable efforts have been made to publish reliable data and information, but the author and publisher cannot assume responsibility for the validity of all materials or the consequences of their use. The authors and publishers have attempted to trace the copyright holders of all material reproduced in this publication and apologize to copyright holders if permission to publish in this form has not been obtained. If any copyright material has not been acknowledged please write and let us know so we may rectify in any future reprint.

Except as permitted under U.S. Copyright Law, no part of this book may be reprinted, reproduced, transmitted, or utilized in any form by any electronic, mechanical, or other means, now known or hereafter invented, including photocopying, microfilming, and recording, or in any information storage or retrieval system, without written permission from the publishers.

For permission to photocopy or use material electronically from this work, please access www. copyright.com (http://www.copyright.com/) or contact the Copyright Clearance Center, Inc. (CCC), 222 Rosewood Drive, Danvers, MA 01923, 978-750-8400. CCC is a not-for-profit organization that provides licenses and registration for a variety of users. For organizations that have been granted a photocopy license by the CCC, a separate system of payment has been arranged.

Trademark Notice: Product or corporate names may be trademarks or registered trademarks, and are used only for identification and explanation without intent to infringe.

A Library of Congress record exists under LC control number:

Publisher's Note
The publisher has gone to great lengths to ensure the quality of this reprint but points out that some imperfections in the original copies may be apparent.

Disclaimer
The publisher has made every effort to trace copyright holders and welcomes correspondence from those they have been unable to contact.

ISBN 13: 978-0-367-25894-8 (hbk)
ISBN 13: 978-0-367-25897-9 (pbk)
ISBN 13: 978-0-429-29042-8 (ebk)

**Visit the Taylor & Francis Web site at http://www.taylorandfrancis.com and the
CRC Press Web site at http://www.crcpress.com**

PREFACE

Extrusion processing has become an important food processing operation. I was introduced to food extrusion when working for General Mills, Inc. starting in 1964. This introduction involved a great deal of empiricism which made it difficult to understand many of the phenomena experienced in food extrusion. At that time, a review of polymer processing literature revealed the strides which had been made in this industry. Through an understanding of rheological and physical properties of the polymers and the hydrodynamic analysis of single-screw extruders, it has been possible to model, optimize, and better control the extrusion process. It has been my desire to see similar strides made in food extrusion despite the added complexities imposed because of the reactive nature of food ingredients and their natural variability.

No text currently exists on food extrusion that brings together the peculiarities associated with foods and a rigorous fundamental approach to the extrusion process. This text and reference book has been written and designed to be used as a self-study and reference book for food engineers and scientists working with extrusion. With this group in mind, the fundamentals of rheology and fluid mechanics applied to food extrusion are developed and reviewed in Volume I. To receive maximum benefit from these developments, it is necessary for the student to understand differential equations. Vector notation was not used in the text to better allow the intended audience to grasp the meaning of the equations. The solutions of the differential equations are, however, given and example problems are intended to illustrate their application. The development of the extrusion theory early is necessary so that it can be helpful in explaining and extending the extrusion studies reported in the literature.

The middle portion of the book is intended for both the novice and practitioner of extrusion. Since many varieties of food extruders exist, the chapter on equipment was developed to give a broad understanding of the machines available. Information on extrusion operations, control, experimentation, and food ingredients used in extrusion is directed toward improved extrusion operations.

The remaining chapters are oriented toward reviewing the technology, literature, and developments in the principal areas of food extrusion. The literature relevant to these areas is very diffuse and considerable time was spent merely gathering the information. As of the publication date, every effort was made to provide a complete bibliography and to summarize these sources in terms of the basic extrusion phenomena, and the physical and chemical principles affecting the extrusion process.

The book can also be used as a text. It would be suitable as a senior course for engineers or a graduate course for engineers or food scientists. The material can be covered effectively in a 3-hr semester course. Students should be encouraged to obtain and read a number of the references given.

Both volumes use SI notation for all units. This may prove cumbersome for some who are only familiar with English engineering units. The decision to use SI was made because most college level course work is being taught with SI, all recent publications are written in SI, and the U.S. industry is moving in this direction. SI also makes a much clearer distinction between force and mass which has been very troublesome to anyone working in the field.

I owe a large debt of gratitude to a number of individuals who provided information and inspiration. Specifically, I wish to thank Mr. Richard Enterline, Mr. Bruce Bain, Mr. Lavonne Wenger, Mr. Charles Manley, and Mr. Maurice Williams for taking time to discuss their extrusion experiences and equipment. Industrial practitioners of food extrusion from Quaker Oats, General Mills, Inc., Ralston Purina, Archers-Daniels-Midland, Frito Lay, etc., have provided insight and a better understanding of the technology. The generosity of many equipment, ingredient, and related firms that provided pictures to illustrate the text is also gratefully acknowledged.

A special thanks goes to Prof. Chiam Mannhiem, Department of Food Engineering and Biotechnology, Technion, Haifa, Israel, who allowed me to work a year in that laboratory and library, where I completed the text. Finally, I wish to express my appreciation to Miss Ruth Korn, Mrs. Elaine Barrar, Prof. Bruce Dale, Mr. Arrun Bansal, and Mr. Ron Tribelhorn who read and helped me edit the final manuscript.

Judson M. Harper
Fort Collins, Colorado
November 1979

THE AUTHOR

Judson M. Harper, Ph.D., P.E., is professor of Agricultural and Chemical Engineering at Colorado State University, Ft. Collins.

Dr. Harper obtained his B.S. degree in Chemical Engineering and his M.S. and Ph.D. in Food Technology from Iowa State University of Science and Technology, Ames, Iowa, in 1958, 1960, and 1963, respectively.

Dr. Harper is a member of the American Institute of Chemical Engineers, the American Society of Agricultural Engineers, the Institute of Food Technologists, the American Association for the Advancement of Science, and the American Society of Engineering Education.

Among awards and honors, he is a member of Tau Beta Pi, Sigma Xi, Gamma Sigma Delta, and Alpha Epsilon. He received the Outstanding Young Alumnus Award from Iowa State University in 1970, and the Distinguished Service Award for Academic Department Heads from Colorado State University in 1977. He was a Fullbright-Hayes Scholar to Israel in 1978 and a Lady Davis Fellow at the Technion, Haifa, Israel, in 1978.

Dr. Harper has been a guest lecturer on food extrusion at national and international symposia around the world, and co-project director on an AID/USDA sponsored project to utilize low-cost extruders to manufacture nutritious foods in developing countries. He has published over 60 research papers in food engineering and in the food service area, and has four patents along with numerous papers, reports, and presentations.

DEDICATION
To my wife, Patricia, and my sons, Jayson, Stuart, and Neal . . .

TABLE OF CONTENTS

Volume I

Volume I

Volume II

Chapter 1

FOOD EXTRUSION

I. INTRODUCTION

Webster's defines the verb to extrude, as "to shape by forcing through a specially designed opening often after a previous heating of the material."[6] Extrusion, therefore, is primarily oriented toward the continuous forming of plastic or soft materials through a die. An extruder is a machine which shapes materials by the process of extrusion.

Several designs are possible for extruders. The simplest is a ram or piston extruder. The principal focus for this book involves screw extruders consisting of flighted screw(s) or worm(s) rotating within a sleeve or barrel. The action of the flights on the screw pushes the platicized material forward and creates the pressure behind the discharge die so that it extrudes through the opening.

Cooking extrusion combines the heating of food products with the act of extrusion to create a cooked and shaped food product. Cooking extrusion can be described as a process whereby moistened, starchy, and/or proteinaceous foods are cooked and worked into a viscous, plastic-like dough. Cooking is accomplished through the application of heat, either directly by steam injection or indirectly through jackets, and by dissipation of the mechanical energy through shearing occurring within the dough. The results of cooking the food ingredients during the extrusion process are the gelatinization of starch, the denaturation of protein, the inactivation of many raw food enzymes which can cause food deterioration during storage, the destruction of naturally occurring toxic substances such as trypsin inhibitors in soybeans, and the diminishing of microbial counts in the final product.

The temperatures reached by the food during cooking extrusion can be quite high (200°C) but the residence time at these elevated temperatures is very short (5 to 10 sec) as illustrated in Figure 1. For this reason, extrusion processes are often called HTST (high-temperature/short-time). They tend to maximize the beneficial effects of heating foods (improved digestibility and instantization [precooking]), while minimizing the detrimental effects (browning, inactivation of vitamins and essential amino acids, production of off-flavors, etc.). Once cooked, the product is forced through a die at the extruder discharge where it expands rapidly with some loss in moisture because of a rapid decrease in pressure. After expansion and cooling/drying, the extrudate develops a rigid structure and maintains a porous texture.

A schematic flow diagram of a cooking extruder process is shown in Figure 2 along with ranges of product moistures and temperatures. In the extrusion processing plant, raw ingredients consisting primarily of cereals and proteinaceous oil seeds are received, cleaned/dehulled, stored, blended, and fed to the extruder. Optional preconditioning with water or steam follows. The extruder cooks and/or forms these ingredients to yield expanded textured products or cooked dough pellets which require further processing. Drying or toasting followed by flavor and/or other ingredient addition completes the typical extrusion process.

II. ADVANTAGES

The principal advantages of the modern food extruder, leading to their expanded role in the food processing industry, were given by Smith.[3] This list, which has been expanded and broadened, is given below.

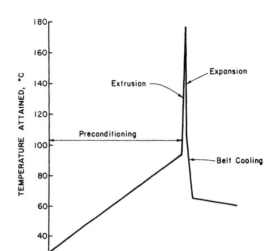

FIGURE 1.   Time-temperature relationship for an extruded food product. (Reprinted from de Muelenaere, H. J. H. and Buzzard, J. L., *Food Technol. (Chicago),* 23, 345, 1969. Copyright© by Institute of Food Technologists. With permission.)

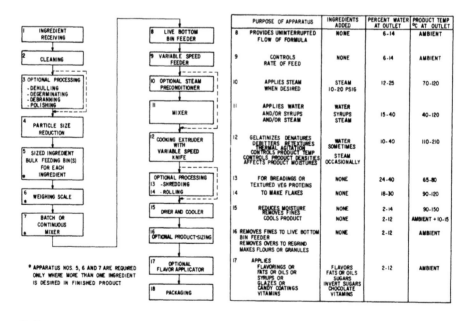

FIGURE 2.   Schematic of cooking extrusion process. (From Smith, O. B., in *New Protein Foods,* Altschul, A. M., Ed., Academic Press, New York, 1976. With permission.)

1.   Versatility — A wide variety of foods can be produced on the same basic extrusion system using numerous ingredients and processing conditions.
2.   High productivity — An extruder provides a continuous processing system having greater production capability than other cooking/forming systems.

3. Low cost — Labor and floor space requirements per unit of production are smaller than for other cooking/forming systems enhancing cost effectiveness.
4. Product shapes — Extruders can produce shapes not easily formed using other production methods.
5. High product quality — The HTST heating process minimizes degradation of food nutrients by heat while improving digestibility by gelatinizing starch and denaturing protein. The short, high-temperature treatment also destroys most undesirable factors in food. Some of these heat denaturable factors are antinutritional compounds such as trypsin inhibitors, hemagglutinins, and gossypol, and undesirable enzymes such as lipases or lipoxidases, microorganisms and other food-borne pests.
6. Energy efficient — Extrusion processing systems operate at relatively low moistures while cooking food products. Lower moisture reduces the quantity of heat required for cooking and redrying the product after cooking.
7. Production of new foods — Extruders can modify vegetable proteins, starches, and other food materials to produce a variety of new food products.
8. No effluents — The lack of process effluents is an important advantage since stringent controls are being placed upon food processors to prevent their releasing pollutants to the environment.

III. HISTORY OF FOOD EXTRUSION

The first food extrusions involved the use of piston or ram-type extruders to stuff casings in the manufacture of sausages and processed meats. The development of the simple food chopper or mincer, consisting of a screw which forces soft food products through a die plate, may have been the first use of a single-screw food extruder. Interestingly, these two early examples of food extruders used by the meat industry exist today in remarkably similar forms. The design of the ram and chopper are relatively simple with recent developments being focused on increased capacity, materials of construction, sanitary design, simplified means of feeding, and improved driving parts such as automated hydraulic activators or electrical motors.

Hydraulically operated batch cylindrical ram macaroni presses came into existence around 1900. They represent the second major food industry to adopt the use of extruders after their initial applications in meat processing.

Rolling or sheeting operations can also be broadly classed as an extrusion process to form cereal dough products which date back to antiquity. The action of two counter-rotating rolls, working toward a restriction or a narrow gap separating them at their nip, effectively shape the dough products while achieving high pressures and shear rates. Rolls have been used to continuously pressurize doughs which are then forced through a nozzle to accomplish forming operations. Ziemba[5] gives several examples where roll-type extruders are applied to form cookies, crackers, confections, marshmallows, and cheese slices.

The initial application of the single-screw extruder, which revolutionized an entire industry, was its use as a continuous pasta press in the mid 1930s. The pasta press mixes semolina flour, water, and other ingredients forming a uniform dough. The screw of the extruder also works the dough and forces the mixture through specially designed dies to create the variety of pasta shapes which are now available. Low shear extrusion screws, characterized by deep screw flights, cause little heat or precooking to occur in the dough.

General Mills, Inc. was the first to use an extruder in the manufacture of ready-to-eat (RTE) cereals in the late 1930s. Initially, RTE cereals only utilized the extruder to shape a precooked cereal dough into a variety of cereal shapes which were dried and subsequently puffed or flaked. Here, the extruder performed a function similar to the pasta extruder by shaping the hot, precooked cereal dough.

Expanded corn collets or curls were first extruded in 1936 but the product was not commercially developed until 1946 by the Adams Corp.® Collet extruders are characterized by an extremely high shear rate within the flights of the screw and grooves in the barrel. No external heat is applied through jackets around the barrel and the entire heat input comes from the viscous dissipation of mechanical energy applied to the shaft of the extruder. Such extruders are often classified as autogenous. The typical collet extruder has its operation limited to low-moisture ingredients and the production of highly expanded snack products. To add to the variety of products made on collet extruders, rice and potato flours as well as a number of starches are used to create numerous shapes including puffs, rings, and French fries.

The desire to precook animal feeds to improve digestibility and palatability led to the development of the cooking extruder in the later 1940s and has greatly expanded the application of extruders in the food field. Cooking extruders come in a variety of sizes and shapes, and provide the capability to vary the screw, barrel, and die configurations for the cooking and shaping requirements of each specific product. To assist in the control of extrusion conditions and temperatures the barrel temperature is controlled using jackets or electrical heating bands. Temperature control is also practiced by direct steam injection into the food in the extruder screw or with hollow screws fitted for circulation of water or steam. Preconditioning the feed materials in an atmospheric or pressurized steam chamber allows the ingredients to be partially cooked and uniformly moistened before entering the extruder. The wide range of moisture contents (10 to 40%), feed ingredients, cooking temperatures (110 to 200°C), and residence times greatly enhance the versatility of the modern day food extruder.

The past three decades have given rise to the ever-increasing number of applications of the cooking extruder. Dry, expanded, extrusion-cooked pet foods quickly developed in the 1950s largely replacing the biscuit baking processes which were used to manufacture them up to that time. The precooking or gelatinization of starches were also areas of additional interest. The grain millers began using an increasing number of extruders to precook cereals for incorporation into nutritious blended foods consisting of a precooked cereal and soy protein in the mid 1960s.

Also in the 1960s, RTE breakfast cereals were developed which were cooked and formed continuously with a one-step process on a cooking extruder. Here, the raw granular cereal and proteinaceous ingredients are added directly to the extruder, cooked by the application of heat through barrel jackets or with steam by direct injection. The high shear environment (created by a shallow-flighted screw rotating in a grooved barrel to prevent slip at the extruder walls) causes the viscous dissipation of a large quantity of mechanical energy. This major source of heat energy enhanced the cooking process by uniformly heating the product from within. The high pressures produced in the cooking extruder are released through special-shaped dies creating an expanded product so the cooking and forming takes place in one machine. Subsequent drying and toasting of the expanded product are required to finish the processing operation.

Applying basic extrusion technology used in the cooking of cereal grains to defatted oil seed protein resulted in a texturized product having meat-like structure and a fibrous texture. This work began in the industrial research laboratories of Archer Daniels Midland, Ralston Purina, and Swift in the early 1960s with the introduction of extrusion-textured protein dough creating the conditions for intramolecular cross-linking necessary for the formation of a fibrous structure.

The ability of the extruder to knead and form products has been further exploited to expand its applicability. A good case is the variable pitch screw extruder which mixes dough in continuous bread-making equipment before it is extruded into pans to be proofed and baked. Other examples can be cited such as the forming and wire cutting of cookie/cracker shapes.

Mechanical deboning of meat has been accomplished with a type of extruder having a conical-shaped screw rotating in a barrel constructed with a supported screen having a plurality of small sharp-edged orifices. An adjustable restriction at the discharge of the deboner results in the screw pressurizing the ground meat and bone mixture. The deboned meat extrudes through the barrel surface and the screen retains the bone fragments which are discharged from the end of the extruder barrel.

Work at the Northern Regional Research Laboratory in conjunction with UNICEF and Wenger Mgf. in the mid 1960s showed that the extruder could be used to produce a well-cooked, full-fat soy flour by denaturing antinuritional factors which exist in the raw bean. The resulting high-fat product is quite stable because the process heat treatment denatured the lipoxidases which normally cause rapid deterioration of fat in the raw soy following grinding.

Pet foods are the largest single product type produced on extrusion equipment. To extend the line of formed pet foods from expanded hard dry varieties made from cereals, oil seed protein, meat and bone meal, fat, and flavorings, semimoist varieties were introduced in the late 1960s. In these formulations, the water activity was reduced in the finished product by adding sugars, propylene glycol, sorbitol, acids and/or other humectants while maintaining final moistures above 30%, consequently the name, "soft-moist". Here, the extruder was used to heat and cook the entire mass and give a final product shape.

Recently, twin-screw extruders have been used to process food products. Two types of twin-screw extruders are employed, one where the two screws rotate the same direction and the other, where the screws rotate in opposite directions. Although these extruders are much more complex than conventional single-screw extruders, they do provide better control of residence time and internal shear of the food ingredients for products that are very heat sensitive. Another advantage is their ability to operate at very low product moistures and product drying is not required.

As a separate issue, it is interesting to observe the development of plastics extrusion equipment in contrast to the food extruder. The initial introduction of single-screw extruders to the forming of thermoplastic resins occurred about the same time as single-screw extruders were introduced to the macaroni industry. Larger, higher capacity extruders with specially designed screws, barrels, and dies have been developed in the plastics industry. Attempts to reduce plastics extrusion from an art to science have been relatively successful and aided these developments. It has been only since 1970 that food extrusion has received the same type of analytical analysis. Although food systems are much more complex and variable than thermoplastics, these research efforts are beginning to pay dividends in basic process understanding and improved designs.

IV. EXTRUDED PRODUCTS

The current variety of extruded food products is impressive.[1,4] Examples of extrusion-cooked products are precooked and modified starches, RTE cereals, snack foods, breading substitutes, beverage bases, soft-moist and dry pet foods, TPP (textured plant protein), full-fat soy flour, soup and gravy bases, and confections. The variations of products within each of these categories greatly expands the list. Many of these variations involve relatively simple changes, such as flavoring or shape. Others involve more fundamental changes such as new ingredient combinations or processing conditions. An example would be the use of two extruders to produce a coextruded product such as a filled tube.

The list of ingredients used in extruded foods consists of almost every imaginable food item. The principal ingredients are all the cereal grains, oil seeds, and legumes. To allow the extruder to tailor the texture and shape of the food to meet specific requirements, modified starches and other hydrocolloids are used. Emulsifiers, fats, and sweetners also have a role in extruded product formulation.

An increasingly important role for food extruders appears assured in the food processing industries because of an expanding interest and demand for fabricated foods.[2] These trends are being driven by an increasing world population and a shrinking conventional energy supply. Both trends focus on the need to transform raw agricultural products consisting of starch, plant protein, and fat directly and efficiently into foods having high acceptability. The ability of the extruder to combine, cook, and texturize food components quickly, continuously, and efficiently makes it ideally suited for this task.

REFERENCES

1. **Sanderude, K. G. and Ziemba, J. V.**, New products come easy with extrusion cooking, *Food Eng.*, 40(8), 84, 1968.
2. **Smith, O. B.**, Extrusion and forming: Creating new foods, *Food Eng.*, 47(7), 48, 1975.
3. **Smith, O. B.**, History and status of specific protein-rich foods. Extrusion-processed cereal foods, in *Protein-Enriched Cereal Foods for World Needs*, Milner, M., Ed., Am. Assoc. Cereal Chem., St. Paul, Minn., 1969, 140.
4. **Williams, M. A.**, Direct extrusion of convenience foods, *Cereal Foods World*, 22(4), 152, 1977.
5. **Ziemba, J. V.**, Extrusion challenges you, *Food Eng.*, 36(8), 57, 1964.
6. *Webster's 3rd New Dictionary*, Merriam Company, Springfield, Mass., 1966, 808.

Chapter 2

THE FOOD EXTRUDER

I. INTRODUCTION

A food extruder consists of a flighted Archimedes screw which rotates in a tightly fitting cylindrical barrel. Raw ingredients are preground and blended before being placed in the feed end of the extrusion screw. In many cases, these ingredients are partially heated and moisturized in a preconditioning chamber which can be of the atmospheric or pressure cooker type. The action of the flights on the screw push the food product forward and in so doing, work and mix the constituents into a viscous dough-like mass. Heat is added to the food dough as it passes through the screw by one or more of three mechanisms: (1) viscous dissipation of mechanical energy being added to the shaft of the screw, (2) heat transfer from steam or electrical heaters surrounding the barrel, and (3) direct injection of steam which is mixed with the dough in the screw.

As the moist, hot food material moves through the extruder, the pressure within the barrel increases due to a restriction at the discharge of the barrel. The restriction is caused by one or more orifices or shaped openings called a die. Discharge pressures typically vary between 30 to 60 atm. At these elevated pressures, boiling or flashing of moisture does not occur within the confines of the barrel because the pressure exceeds the vapor pressure of water at the extrusion temperature. Once the food emerges from the die, the pressure is released causing the product to expand with the extensive flashing of moisture. The loss of moisture from the product results in adiabatic cooling of the food materials with the product reaching a temperature of approximately 80°C in a matter of seconds, where it solidifies and sets often retaining its expanded shape.

Because the flights on the screw of a food extruder are completely full, the food product is subjected to high shear rates as it is conveyed and flows by the action of the screw. These high shear rate areas tend to align long molecules in the food constituents giving rise to cross-linking or restructuring resulting in the extruded foods unique texture. The shear environment in the die assembly of the extruder can also have similar effects.

II. EXTRUDER COMPONENTS

Extruders are unique pieces of food processing equipment. Consequently, it is important that the nomenclature of extruders be carefully examined. In defining the operating parts and features, a more thorough understanding of the operation of the extruder can be achieved, and the basis for further discussion of food extrusion developed. Much of the nomenclature and terminology used in food extrusion borrows heavily from plasticating extrusion processes and technology.

A. Extrusion Drive

Figure 1 shows a typical food extruder and the major components of the stand, drive, and transmission as defined below.

1. Support or Stand

The extruder is mounted on a steel or cast iron frame which can be bolted to the floor. The frame positions the working members of the extruder so that they can func-

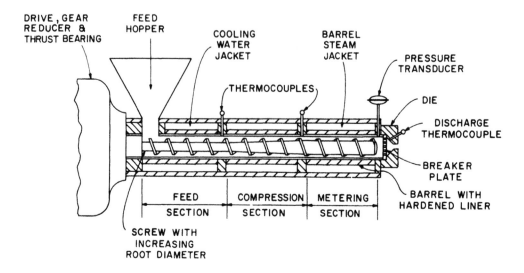

FIGURE 1.   Cross-section of a typical food extruder. (Reprinted from Harper, J. M., *Food Technol. (Chicago)*, 32(7), 67, 1978. Copyright © by Institute of Food Technologists. With permission.)

tion easily, and supports the major extruder components including feeders, screw, barrel, and drive.

2. Drive Motor

Electric motors are usually used to drive food extruders. The size of the motor depends on the capacity of the extruder and may be as large as 300 kW (400 hp).

3. Speed Variation

Variation of the screw speed on the extruder is a valuable control parameter. Continuous variations are provided by magnetic, electrical or mechanical means. Because of this need for continuous control of speeds, without mechanical components which wear and give imprecise control over time, SCR (silicon-controlled rectifier) DC motors are becoming widely accepted.

4. Transmission

The screw speed on food extruders is normally less than 500 rpm. A transmission is used to reduce the speed with a proportional increase in the torque of the drive motor. Belts or chains are sometimes used as reducers but gear reducers with positive lubrication are more frequently employed for the high loads encountered in extrusion.

5. Thrust Bearing

A bearing is required to support and center the extrusion screw and absorb the thrust exerted by the screw. Since the screw forces food forward against a back pressure, a rearward thrust is produced on the screw which must be absorbed by the thrust bearing. The magnitude of the thrust is approximated by the maximum extrusion pressure times the cross-sectional area of the barrel. The thrust bearing must be able to sustain the load produced under normal extrusion conditions giving an expected life of 20,000 to 50,000 hr.

B. Feed Assembly

The feeding, blending, and preconditioning, which consists of moisturizing and/or heating of ingredients, is an essential part of an extrusion operation. The consistent and uniform feeding of food and other ingredients is necessary for the consistent operation of an extruder.

1. Hoppers or Bins

Steel bins are used to hold food ingredients above feeders. Bin vibrators are often required to aid the movement of the ingredients into the feeders and prevent bridging. These vibrators can be either oscillators, jets of air, or internally mounted screw conveyors that move food materials into the discharge throat of the hopper or bin. Hoppers, equipped with devices at their discharges to assure a continuous flow of ingredients, are said to have "live-bottoms".

2. Feeder (Dry Ingredient)

A device providing a uniform delivery of food ingredients which are often sticky, non-free-flowing substances. A number of feeding devices are used including the following. The relative costs and complexity of the feeders are in the order which they are listed.

Vibratory feeders — A vibrating pan which has either variable frequency or stroke and is used to control the feed rate from a hopper.

Variable speed auger — Screw conveyors having a variable speed auger can be used to volumetrically meter relatively free-flowing food ingredients.

Weigh belts — A moving belt with a weighing device under the central section of the belt is used to gravimetrically meter food ingredients. Two types of weigh belts exist — one with a variable feed gate opening to the belt and the other having a variable speed belt with a fixed gate. Signals from the weighing device adjust either the gate opening or belt speed to accomplish a uniform feed rate.

3. Slurry Tanks

Storage tanks for many liquid ingredients or slurries which are added to the extruder are called slurry tanks. Agitators and heating jackets are often required for slurry tanks to keep ingredients heated, uniformly dispersed, and free flowing.

4. Liquid Feeders

The uniform metering of liquid ingredients is accomplished in a variety of ways.

Water wheels — A rotating wheel with a plurality of cups attached to its periphery can be used as a metering device. The level of liquid in the reservoir into which the cups dip must be constant. The discharge rate is proportional to the speed of rotation of the wheel.

Positive displacement metering pumps — Liquids are metered at a constant rate by a positive displacement metering pump whose length of stroke or speed of rotation can be varied.

Variable orifice — A needle valve is used to create a variable orifice opening to control liquid flow. Automatic positioning of the needle valve can be accomplished by a diaphragm activator whose position is established by a feedback controller that operates on a signal, proportional to the flow rate measured by a primary sensor. In such systems, the pressure of the liquid stream behind the control valve is maintained constant by a regulator.

Variable head — The pressure, or more normally, the height of liquid entering a fixed orifice can be used to control liquid flow rate.

Liquid flow control systems are classified as either feedforward or feedback systems. Feedforward systems operate on the basis that a setting will give a fixed and constant feed-rate response, while feedback systems adjust the position of the primary control element on the basis of a measurement of the actual flow which is fed back to the primary control element.

5. Batch Feed System

The proper ratio of individual feed ingredients is maintained by weighing them into a large weigh hopper on a batch basis. This batch of ingredients is mixed and fed as a preblend to the extruder.

6. Continuous Feed Systems

Individual ingredient streams are continuously combined in the proper proportion or ratio. Individual feeders on each stream control the rate of addition of the various ingredient types to a continuous blender before they enter the extruder.

7. Preconditioner

A closed vessel where the feed ingredients are mixed with water, steam, and/or other liquid ingredients for the purpose of increasing their moisture and/or temperature. In some cases, the preconditioner can serve as the continuous blender for several individual feed streams. The preconditioner can be designed and operated as either a pressure or atmospheric vessel. When it functions under pressure, a rotary valve or similar functioning device must be used to maintain the pressure differential between the preconditioner and the surroundings. A variety of mixing devices are used within the preconditioner.

Ribbons — Helical ribbon flights attached to the central shaft of the preconditioner. The pitch of the ribbon flights serve to convey the feed forward, within the preconditioner, toward the extruder inlet.

Paddles — Small, flat-bladed paddles can extend from the central preconditioner shaft(s). The angle the blades make with the drive shaft is often adjustable to alter the conveying action within the preconditioner.

8. Rotary or Star Valve

Used as a continuous feed port which maintains a seal between the internal pressure of the preconditioner and the external atmospheric pressure. The rotary valve consists of a driven rotating wheel with deep pockets which fit tightly within a valve body designed with opposing feed and discharge ports. Feed material falls into a pocket under the feed port and by the rotation of the wheel, are transported to the discharge port where they fall out. The vanes on the rotating wheel serve as the physical barrier between the high and low pressure sides of the device.

9. Feed Transition

A device which accepts the feed ingredients directly from the preconditioner. In the case of free-flowing ingredients, this may be as simple as an open hopper. In instances where the ingredients are sticky, a forced feeder consisting of a feed screw or hopper with paddles may be used. To maintain a pressure difference between the preconditioner and the feed port on the extrusion screw, the feed transition may incorporate a rotary valve or other pressure differentiating device.

C. Extrusion Screw

The screw is the central portion of a food extruder. It accepts the feed ingredients at the feed port, conveys, works, and forces them through the die restriction at the discharge. The extrusion screw is divided into three sections (as shown in Figure 1) whose names correspond to the function each section plays.

1. Feed Section

The portion of the screw which accepts the food materials at the feed port or throat. Usually the feed section is characterized by deep flights so that the product can easily fall into the flights. The function of the feed section is to assure sufficient material is moved or conveyed down the screw and the screw is completely filled. When a screw is only partially filled with feed materials, the condition is termed starved feeding. The feed section typically is 10 to 25% of the total length of the screw.

2. Compression Section

The portion of the screw between the feed section and the metering section. Sometimes the transition section is called the compression section. Compression is achieved in several ways, as shown in Figure 2. The most common way being the gradual decrease in the flight depth in the direction of discharge. Another is a decrease in the pitch in the transition section.

The food ingredients are normally heated and worked into a continuous dough mass during passage through the transition section. The character of the feed materials changes from a granular or particulate state to an amorphous or plasticized dough. It is common for the transition section to be the longest portion of the screw and approximately half its length.

3. Metering Section

The portion of the screw nearest the discharge of the extruder which is normally characterized by having very shallow flights. The shallow flights increase the shear rate in the channel to the maximum level within the screw. The viscous dissipation of mechanical energy is typically large in the metering section so that the temperature increases rapidly. The high shear rate in the screw also enhances internal mixing to produce temperature uniformity in the extrudate. In some cases, pins or cut flights are employed to increase mixing and mechanical energy dissipation. The critical screw dimensions are shown in the sectional drawing of an extrusion screw in Figure 3. The dimensions, terminology, and their interrelationships are described below.

Flight — The helical metal rib wrapped around the screw whose action on the food mechanically advances it toward the discharge of the screw. In most instances, the flight is continuous but when it is broken, it is called a cut flight.

Root — The continuous central shaft of the screw which is usually of cylindrical or conical shape.

Land — The surface at the radial extremity of the flight which constitutes the outside periphery of the screw. In some cases the screw land is especially hardened by flame induction or by depositing a hard facing metal surface.

Leading flight edge — The pushing face of the flight which faces the extruder discarge. Wear on the flight is seen first on the leading flight edge.

Trailing flight edge — The face of the flight which faces toward the discharge of the extruder.

Flight shape — Flights are usually rectangular or trapezoidal in shape. The trapezoidal shape results in a channel which is wider at the surface of the screw than at the root, thus reducing stagnation of product in the corners between the flight and screw root.

Screw shank — The rear protruding portion of the screw to which the extruder drive is attached. Normally the shank passes through a seal in the support stand before the thrust bearing, which prevents leakage of food materials or grease. The shank is fitted with a keyway which mates with a drive coupling. Soft keys or shear pins can also serve as overload protection devices.

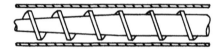

I. INCREASING ROOT DIAMETER

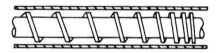

2. DECREASING PITCH, CONSTANT ROOT
   DIAMETER

3. CONSTANT ROOT DIAMETER SCREW
   IN BARREL WITH DECREASING
   DIAMETER

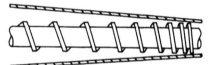

4. CONSTANT ROOT DIAMETER, DECREASING
   PITCH SCREW IN BARREL WITH
   DECREASING DIAMETER

5. CONSTANT ROOT DIAMETER, CONSTANT
   PITCH SCREW WITH RESTRICTIONS
   IN CONSTANT DIAMETER BARREL

FIGURE 2.   Varying configuration of screw and barrel to achieve compression. (From Harper, J. M., *Crit. Rev. Food Sci. Nutr.*, 11(2), 155, 1979. With permission.)

Screw diameter, D, — Nominally the internal diameter of the barrel, in which the screw rotates, is denoted as D. The actual screw diameter accounting for the clearance, δ, is

$$D_s = D - 2\delta \qquad (2.1)$$

Flight height, H, — Nominally the dimension between the barrel surface and root of the screw is defined as H. The actual flight height of the screw is

$$H_s = H - \delta \qquad (2.2)$$

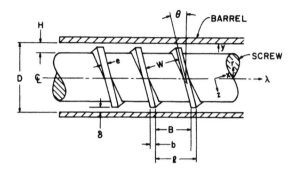

FIGURE 3.   Geometry of the metering section of an extrusion screw. (From Harper, J. M., *Crit. Rev. Food Sci. Nutr.*, 11(2), 155, 1979. With permission.)

Root diameter, D_r — Diameter of the root of the screw.

$$D_r = D - 2H = D_s - 2H_s \qquad (2.3)$$

Diametral screw clearance, 2δ — The difference between the diameter of the screw and the bore of the barrel.

$$2\delta = D - D_s \qquad (2.4)$$

Radial screw clearance, δ — Half the diametral clearance.

Lead, l — The axial distance from the leading edge of a flight at its outside diameter to the leading edge of the same flight in front of it.

Helix angle, θ — The angle which the flight makes with a plane normal to the axis of the screw.

$$\theta = \tan^{-1} \frac{l}{\pi D_s} \approx \tan^{-1} \frac{l}{\pi D} \qquad (2.5)$$

Channel — The helical opening which extends from the feed to discharge end of the screw. The channel is formed on two sides by screw flights, on the bottom by the screw root, and enclosed in the top by the inner barrel surface.

Axial channel width, B — The axial distance from the leading edge of one flight to the trailing edge of the same flight one complete turn away at the diameter of the screw.

Channel width, W — The channel width measured as above but perpendicular to the flight.

$$W = B \cos \theta \qquad (2.6)$$

Axial flight width, b — The axial width of a flight measured at the diameter of the screw.

$$b = l - B \qquad (2.7)$$

Flight width, e — The flight width measured perpendicular to the face of the flight.

$$e = b \cos \theta \qquad (2.8)$$

Reference axes, x, y, z — The following directions are adopted by convention: x — cross-channel direction, z — down the channel direction along the screw helix, y — direction perpendicular to root of screw. λ is in the direction of the axis of the screw or along its length.

Channel length, Z — The length of the screw channel in the z direction. The length of channel in one complete turn of the helix is given by:

$$Z = \ell/\sin \theta \qquad (2.9)$$

Peripheral speed or tip velocity, V — The speed of the screw tip.

$$V = \pi D_s N \approx \pi D N \qquad (2.10)$$

Number of flight turns, p — Total number of single flights in an axial direction.

Single flighted screw — A screw having a single helical flight. In this case, the parameter p = 1.

Multiple flighted screw — A screw having more than one helical flight in parallel such as a double flighted screw where p = 2. Viewed from the discharge tip or nose of the screw, two flights can be seen terminating.

Axial area of screw channel — The cross-sectional area of the channel measured in a plane through the axis of the screw.

Developed volume of screw channel — The volume developed by the axial area of the screw channel with one revolution of the screw axis.

Compression ratio, C.R. — The factor obtained by dividing the developed volume of the screw channel at the feed opening by the developed volume of the last full flight prior to discharge. Sometimes C.R. is incorrectly given as the ratio of H in the feed section to H in the metering section which is really the channel depth ratio. Typical ranges in C.R. are from 1 to 5:1.

Height to diameter ratio, H/D — The ratio of the flight height to screw diameter normally calculated for the metering zone.

Hollow screw — Sometimes extrusion screws have a hollow internal core to accept either a heating or cooling medium thus providing extra heat transfer surface area.

Nose — The discharge tip of the extruder screw which can be pointed or blunt depending upon the design of the die head assembly.

D. Extruder Barrel

The extruder barrel is the cylindrical member which fits tightly around the rotating extruder screw. The cross section of a segment of an extruder barrel is shown in Figure 4. The nomenclature and features of the barrel are discussed below.

1. Bore, D

The inside diameter of the barrel and the nominal size of the extruder. Internal diameters range from 5 cm to 25.4 cm.

2. Length to Diameter Ratio, L/D

The distance from the rear edge of the feed opening to the discharge end of the barrel bore divided by the bore diameter to express a ratio where the diameter is reduced to 1. Food extruders typically have L/Ds ranging from 1:1 to 20:1.

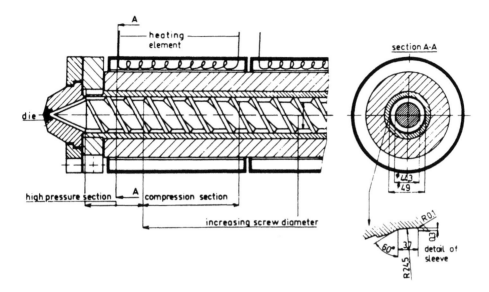

FIGURE 4. Cross-section of extruder barrel with detail of grooves in barrel sleeve. (From Van Zuilichem, D. M. and Stolp, W., Developments and applications of starch-based ingredients in the manufacture of extruded foods, Int. Snack Seminar at the Central College of the German Confectionery Inst., Solingen, Germany, October 18 to 21, 1976. With permission.)

3. Segmented Barrel

A barrel made up of segments. The barrel is fabricated by clamping or bolting together several segments. Such an arrangement makes it relatively easy to alter the interior conformation of the barrel and to replace the discharge section which wears the most rapidly. Segmented barrels have the disadvantage that they tend to leak between the segments if not carefully assembled.

4. Feed Throat or Opening

A hole through the feed end of the barrel for the introduction of feed material usually at least one diameter in length. The shape of the opening can vary as shown in Figure 5.

Vertical — The feed throat has vertical sides which are tangent to the sides of the bore.

Sloped — The side of the throat toward which the screw rotates is inclined. Such a design improves the manner in which feed materials are accepted into the screw.

Undercut — The area open to feed is cut under the screw and improves the feeding of soft or rubbery-like materials.

Side — An opening which feeds the material into the side of the screw.

5. Barrel Liner

A removable sleeve within the barrel. Usually, the liner is of specially hardened or cast materials which resist wear.

6. Hardened Material of Construction

Barrels are constructed of special hard alloys such as Xaloy® 306 and stainless steel 431 (heat treatable). Nitriting is also used as a method to produce a very hard surfaced material which is wear resistant.

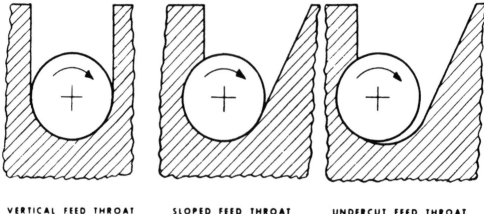

VERTICAL FEED THROAT   SLOPED FEED THROAT   UNDERCUT FEED THROAT

FIGURE 5.   Various extruder feed throats. (From Paton, J. B., Squires, P. H., Darnell, W. H., and Cash, F. M., in *Processing of Thermoplastic Materials,* Bernhardt, E. C., Ed., Robert E. Krieger, Huntington, 1974, 157. With permission. © 1959 Society of Plastics Engineers, Inc.)

7. Grooves or Splines

The interior surface of the barrels of food extruders often have small grooves to prevent slippage of the foot material at the walls. The presence of grooves increases the ability of the extruder to pump food materials against high back pressures. The grooves are normally shallow and can take on a variety of configurations.

Straight — Grooves which run axially down the length of the barrel.

Spiral — Helical grooves inside the bore which normally have a hand opposite that of the screw.

8. Vent

An opening at an intermediate point in the extruder barrel to permit the removal of air, steam, or other volatile materials from the material being processed. Vented extruders tend to act like two extruders placed in series.

9. Jacket

Hollow cavities around the outside of the barrel in which a heat transfer medium can be circulated such as water, oil, or steam. To enhance heat transfer, fins are often cast into the outer surface of the barrel. The fins have the ability to increase the outside heat transfer surface of the barrel and the velocity of the circulating heat transfer media.

10. Heaters

Under conditions when only heating is required, a variety of heating elements can be placed directly around the barrel to accomplish this task. Electrical resistance band heaters are common but tubular resistance heaters or induction coils have also been used.

E. Extruder Discharge

Once the food material leaves the end of the extrusion screw, it enters the discharge section of the extruder which normally holds the extruder die, cutters, and take-away devices. Each of these important elements is discussed below.

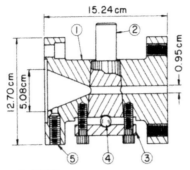

1. BODY
2. PLUG
3. RETAINER PLATE
4. ½ in. DIAMETER HARDENED BALL
5. STOCK THERMOCOUPLE OR
   PRESSURE GAUGE FITTING

FIGURE 6. Gate assembly valve for extruder discharge. (From Paton, J. B., Squires, P. H., Darnell, W. H., and Cash, F. M., in *Processing Thermoplastic Materials*, Bernhardt, E. C., Ed., Robert E. Krieger, Huntington, 1974, 247. With permission. © 1959 Society of Plastics Engineers, Inc.)

1. Die Head Assembly

Holder for the extrusion die, this discharge valve sometimes serves as the support for the cutter.

2. Breaker Plate

This perforated plate located at the end of the extruder can serve as a mechanical seal between the barrel lining and die, and also provides a more even pressure distribution behind the die plate. The thickness of the breaker plate is normally 20% of the barrel diameter and contains a number of counter-sunk holes ranging from 3 to 6 mm in diameter. In some cases, screens called a screen pack are placed on the upstream side of the breaker plate which acts as their support. The function of the screen pack is to remove large or uncooked chunks of food which might otherwise clog the small openings in the die.

3. Gate Assembly

In certain instances, an assembly can be attached directly to the discharge of the extruder which contains an adjustable discharge valve. An example is shown in Figure 6. Gate assemblies are designed for pressures greater than 400 atm.

4. Dead Volume

This refers to the volume of the die head assembly. If the dead volume is large, then uneven temperatures can occur within the head causing nonuniform flow through the holes of the die.

5. Die

Extrusion dies are small openings which shape the food material as it flows out of

FIGURE 7.   Die plate with holes for die inserts. Central face cutter in position. (Courtesy of Anderson-Ibec, Strongsville, Ohio. With permission.)

the extruder. The shape of the dies vary, the simplest being a hole. Annular openings and slits are also common. Often the opening on the feed side of the die is streamlined to improve uniformity of the extrudate.

6. Die Plate

This heavy plate contains numerous holes which can receive individual die inserts containing the actual die opening. Should one hole become damaged, an individual die insert can be replaced rather than the entire die. Therefore, use of the plate/insert approach is quite popular. An example is shown in Figure 7.

7. Die Insert

Individual die openings exist as inserts which slip into holes in a die plate. The insert is positioned in the die plate from the rear so its collar positions and holds the insert in place. Special Teflon® coatings are being used to reduce dough drag in the actual die opening.

8. Quick Change Dies or Breaker Plates

A new die plate or breaker plate can be quickly slipped into place by means of a hydraulic piston when the original die plate ceases to operate properly or becomes clogged. The quick change operation can be likened to changing a slide in a projector. Using the quick change device, little production time is lost.

9. Cutters

A common cutter used with extruder operations is called a face cutter, meaning the cutting knives revolve in a plane parallel to the face of the die. The relative speed of rotation of the cutter knives and the lineal speed of extrudate results in the control of the length of the extruded pieces. Depending upon the location of the cutter drive motor and the length of the arms holding the cutting knives or blades, two different types of cutters exist.

Fly cutters — Motors are located outside the central axis of the die. Relatively long cutter arms rotate with very high tip velocities.

Central cutter — Cutter blades rotate about the central axis of the die assembly, are relatively short, and operate at a lower tip velocity than do those on fly cutters.

10. Take-Away

Once the extruded pieces are cut, they normally fall into some type of take-away system. This can be pneumatic or mechanical when small pieces are involved. Care must be exercised in the design of the take-away system to avoid clumping of the sticky and hot extruded pieces. Extruded sheet goods are removed from the extruder by a series of driven rolls which serve as the take-away system.

Chapter 3

DOUGH RHEOLOGY

I. INTRODUCTION

Food extrusion is a flow process requiring the characterization of food dough rheology for the description of many important extrusion parameters. Some of these extrusion parameters are

1. Velocity profiles in the extrusion screw channel leading to defining the extrusion rate
2. Energy requirements for the extrusion process
3. Pressure profiles along the extrusion screw
4. Residence time distributions of the food materials passing through the extruder
5. Flows and pressure drops through extrusion dies
6. Heat transfer between the barrel and food dough

This chapter will cover the fundamentals of rheology as they pertain to low and intermediate moisture food doughs. Some underlying definitions and concepts of rheology will be considered first, followed by a summarization of existing information on food dough rheology. Last, there will be a discussion of measurements techniques used to determine the rheology of food doughs.

II. RHEOLOGY

Rheology can be defined as the science of the deformation and flow of matter. In extrusion, the behavior of the fluid food dough under the influence of a shearing stress is of primary interest. Figure 1 describes an idealized situation, giving rise to the definition of viscosity and the underlying factors which distinguish various rheological behaviors. Two parallel plates are separated by a distance Y and filled with an ideal fluid with no leakage at the edges. A shearing stress, τ_{yx}, is placed on the fluid by the shear force, F, which causes the lower plate of area, A, to move at a constant velocity, V_x. The subscripts, yx, on the shear stress indicate that the stress occurs perpendicular to the y axis in the x direction. A linear velocity gradient in the fluid between the two plates results and there is no slip at the wall. The slope of the gradient is $-V_x/Y$ and is termed shear rate. The relationship can be written

$$\frac{F}{A} = -\mu \frac{V_x}{Y} \tag{3.1}$$

or

$$\tau_{yx} = -\mu \frac{dv_x}{dy} \tag{3.2}$$

which states that the shear force per unit area is directly proportional to the negative velocity gradient or shear rate. Equation 3.2 is Newton's law of viscosity and fluids which behave in this manner are called Newtonian fluids. The negative sign is necessary to obtain a positive τ when shear rate is negative, thus conforming with the normal sign convention.

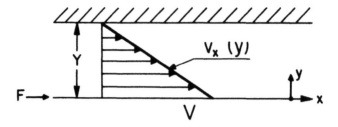

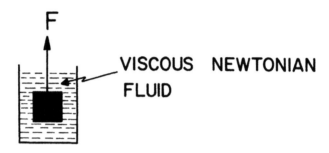

FIGURE 1.   Idealized fluid shear between two parallel plates.

FIGURE 2.   Idealized Newtonian dashpot.

When the direction of the shear stress and the shear rate is obvious, the negative sign is usually ignored and Equation 3.2 is rewritten as

$$\tau = \mu\dot{\gamma} \tag{3.3}$$

where shear rate, dv_x/dy, is the time derivative of the strain, $dx/dy = \gamma$, placed on the system. Units of shear rate are s^{-1} and shear stress are N/m^2 making the units on viscosity Ns/m^2.

The behavior of a Newtonian fluid can be easily visualized when considering a dashpot filled with a Newtonian fluid as shown in Figure 2. When a force, F, is applied to the piston in the dashpot, the piston moves through the Newtonian fluid with a constant velocity. The velocity is directly proportional to the force applied if there is no turbulence and the gravitational and inertial forces and end effects of the piston are ignored. When F is removed, the piston stops moving immediately and will not return to its original position. The dashpot then exhibits ideal Newtonian behavior or a linear relationship between shear stress and shear rate with no memory.

All gases and many simple fluids obey closely Newton's law of viscosity. Since most foods consist of biopolymeric materials, such as starches and proteins, nonideal or non-Newtonian flow behavior results and this must be considered.

III. NON-NEWTONIAN BEHAVIOR

Several types of non-Newtonian viscous behaviors have been recognized. These different types of behavior can be recognized by plotting τ vs $\dot{\gamma}$ as has been done in Figure 3. First, classical Newtonian behavior is characterized by the linear relationship between τ and $\dot{\gamma}$. The slope of the line is constant and equal to μ. Bingham plastics exhibit a yield stress, τ_0, before they begin to shear, but once flow begins they behave as a Newtonian fluid. The simple model

$$\tau = \mu\dot{\gamma} + \tau_0 \tag{3.4}$$

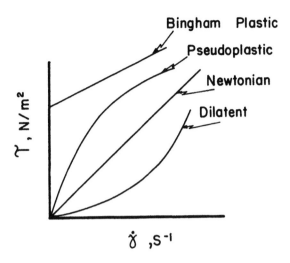

FIGURE 3. The relationship between τ and $\dot{\gamma}$ for several types of viscous behavior.

adequately describes the shear stress-shear rate relationship for the Bingham plastic.

Two classical, non-Newtonian fluid behaviors are shear thinning and shear thickening. Pseudoplastic fluids show a reduced apparent viscosity as the shear rate increases while dilatent fluids exhibit an increased viscosity with increasing $\dot{\gamma}$. A variety of rheological models have been proposed to describe these types of non-Newtonian behaviors as discussed in detail by Holdsworth.[8] Because of its simplicity and ability to correlate observed rheological behavior over the shear rate range of 10 to 200 s^{-1} found in food extruders (see Rossen and Miller[18]), the Ostwalde and de Waele or power law model has found extensive use as a constitutive equation describing food doughs.

The power law model can be written as

$$\tau = m\dot{\gamma}^n \tag{3.5}$$

where τ = shear stress, N/m^2, m = consistency index, Ns^n/m^2, n = flow behavior index, and $\dot{\gamma}$ = shear rate, s^{-1}. It states that the shear stress is proportional to the shear rate raised to the power n. For pseudoplastic fluids n < 1 and for dilatent fluids n > 1. The degree of the nonideal fluid behavior exhibited by a fluid can be measured by the extent which the flow behavior index, n, deviates from 1. When n = 1, the power law simply reduces the Newton's law of viscosity where the consistency index is equal to the viscosity, μ. The power law model has also been written with an additional term accounting for a yield stress. Such a form is called the Herschel-Buckley model.

Clark[4] has discussed the reasons for the acceptance of the power law model (for describing the rheological behavior of intermediate and low moisture food doughs); it fits experimental data reasonably well, is convenient to use, and handles both pseudoplastic and dilatent behavior. Despite these advantages, the simple power law has certain limitations which should also be mentioned. It does not account for a yield stress, does not extrapolate well to shear rates outside the range of the data, and can be applied over only a limited range of $\dot{\gamma}$. Yield stresses are not particularly important in extrusion work and have only been found with a limited number of food doughs.

Figure 4 shows the typical τ vs. $\dot{\gamma}$ curve for a pseudoplastic material over a wide range of $\dot{\gamma}$ and illustrates the limitation of the power law model in fitting the actual data at both very low or very high $\dot{\gamma}$. Examining the curve shows that a pseudoplastic

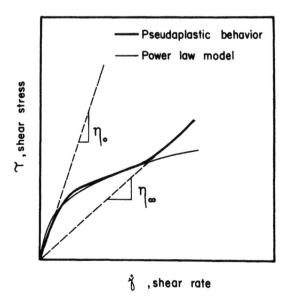

FIGURE 4. Typical pseudoplastic fluid behavior over a
wide range of shear rate (Reprinted from Clark, J. P., *Food
Technol. (Chicago)*, 32(7), 73, 1978. Copyright © by Institute of Food Technologists. With permission.)

behaves like a Newtonian fluid at both very low and very high shear rates. Conse-
quently, straight lines having a slope of η_0 at low $\dot{\gamma}$ and $\eta\infty$ at high $\dot{\gamma}$ fit the data very
well in both these regions. The Reiner-Philippoff model

$$\tau = (\eta_\infty + \frac{1 + (\tau/\tau_s)^2}{\eta_0 - \eta_\infty})(-\dot{\gamma}) \tag{3.6}$$

contains three constants, η_0, $\eta\infty$, and τ,, and can fit experimental data over a much
wider range of $\dot{\gamma}$. The power law is unable to correlate τ and $\dot{\gamma}$ data at high shear rates
and only crudely approximates the response at low shear rates. Since extrusion proc-
esses involve about a 20-fold range of shear rate, the fact that the power law model is
only good for a limited range is not a particularly severe disadvantage; it is oversha-
dowed by the advantage of having to determine only two parameters.

The existence of an apparent Newtonian shear stress-shear rate relationship for pseu-
doplastics at very low and high $\dot{\gamma}$ has been attributed to the nature of the molecules in
the food doughs. Most food doughs consist of biopolymers of starch or protein which
are relatively long and large molecules. At low shear rates, these molecules have a
random orientation so that the entire fluid body shears and momentum is transferred
uniformly from fast to slow moving elements giving rise to an apparent Newtonian
behavior. As the shear rate increases, the long biopolymeric molecules tend to align
themselves along the streamlines of flow resulting in a reduced shear stress with increasing shear rate. At very high shear rates, the molecules are completely oriented
and the relative motion between molecules is primarily influenced by the shear rate
rather than random Brownian motion at the molecular level. In the aligned configuration, the fast moving molecules transfer momentum to slower moving molecules by
some sort of drag process. Under these circumstances, it would be expected that the
Newtonian viscosity exhibited by a pseudoplastic at high $\dot{\gamma}$ would be lower than that
exhibited at low $\dot{\gamma}$.

A. Apparent Viscosity

Once the correlations between τ and $\dot{\gamma}$ have been considered, it is necessary to discuss apparent viscosity. Apparent viscosity is the viscosity which exists at a particular value of $\dot{\gamma}$ and changes with differing values of $\dot{\gamma}$ for non-Newtonian fluids. Simply, apparent viscosity is defined as

$$\eta = \tau/\dot{\gamma} \ . \tag{3.7}$$

The symbol η is used to depict apparent viscosity and clearly distinguishes it from μ which is a fluid property for Newtonian fluids. Substituting for τ in terms of $\dot{\gamma}$ using Equation 3.5 yields

$$\eta = m\dot{\gamma}^n/\dot{\gamma} = m\dot{\gamma}^{n-1} \tag{3.8}$$

where η = apparent viscosity, Ns/m^2, m = consistency index, Ns^n/m^2, $\dot{\gamma}$ = shear rate, s^{-1}, and n = flow behavior index. Since the negative sign has been dropped out of Equation 3.8, the user has to treat the shear rates as positive instead of rigorously preserving the sign convention. Making a log-log plot of Equation 3.8 results in a straight line, with a slope of n − 1. Such a plot is given in Figure 5 which also shows a plot of log τ vs log $\dot{\gamma}$ which has a slope of n as indicated by Equation 3.5.

B. Rheological Model

The effect of moisture content and temperature has been included in models for apparent viscosity. Since both the temperature and moisture of the food dough are often changed as a part of the control and operation of a food extruder, it is important to understand how these variables effect the apparent viscosity. The Arrhenius equation form has been used to correlate changes in temperature for relatively high-temperature, low-moisture doughs.

$$\eta = \eta_1 \exp(\Delta E_n/RT) \tag{3.9}$$

where η = apparent viscosity at temperature T, Ns/m^2, η_1 = apparent viscosity at a high temperature, Ns/m^2, ΔE_n = energy of activation for flow, J/g mol, R = gas constant, 8.314 J/g mol °K, and T = absolute temperature, °K. Equation 3.9 shows changes in the apparent viscosity to be proportional to the exponential of reciprocal absolute temperature.

To account for changes in apparent viscosity with changes in moisture, a logarithmic mixing role form has been employed which can be simply given as

$$\eta = \eta_2 \exp(KM) \tag{3.10}$$

where η = apparent viscosity at moisture, M, Ns/m^2, η_2 = viscosity at reference moisture content, Ns/m^2, K = constant, and M = moisture. Such an equation form assumes no interaction between moisture and the other molecular species in the food dough and does not extrapolate to the correct viscosity at 100% moisture. Therefore, it appears best suited for correlations where deviations in moisture are small, and for fully cooked doughs where moisture does not combine further with the food ingredients.

Harper, Rhodes, and Wanninger[7] combined Equations 3.8, 3.9, and 3.10 logarithmically to give

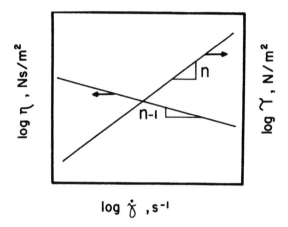

FIGURE 5.   Log-log plot of η vs $\dot{\gamma}$ and τ vs $\dot{\gamma}$.

$$\eta = \eta^* \dot{\gamma}^{n-1} \exp(\Delta E_\eta/RT) \exp(KM) \qquad (3.11)$$

where η^* = reference apparent viscosity, Ns/m², and all other symbols have been previously described. This model requires estimating four parameters using a least squares regression analysis which can be most easily done when the equation is linearized by taking the log of both sides.

Although Equation 3.11 accounts for some important factors which affect dough viscosity, it clearly does not account for changes in apparent viscosity which occur due to the cooking of the food constituents in the dough. Two types of cooking reactions which occur with food biopolymers are protein denaturization and starch gelatinization. In these reactions, water and the food materials themselves interact to create new and altered forms which have a distinctly different rheological behavior. The cooking process is certainly time and temperature dependent but probably changes with the concentration and quantities of the chemical species present, and the shear environment. The chemical reactions which occur also have heats of reaction which are supplied or removed during the cooking extrusion process. Food extrusion results in the chemical alteration of the feed ingredients through the cooking and texturizing process, and in this respect is totally different from the melting processes which occur during the extrusion of thermoplastic resins.

To model the time-temperature effects on the apparent viscosity of soy protein doughs, Remsen and Clark[17] applied the work of Roller,[20] who characterized the time-temperature effects on apparent viscosity during the curing of epoxy resins. Their model can be written for constant moisture doughs as

$$\eta = \eta^* \dot{\gamma}^{n-1} \exp(\Delta E_\eta/RT) \exp \int_{t_o}^{t} k_\infty \exp(\Delta E_k/RT) dt \qquad (3.12A)$$

where k_∞ = apparent kinetic factor at infinite t, min⁻¹, ΔE_k = activation energy for cooking reaction, J/g mol, R = gas constant, 8.314 J/g mol °K, t_o = time at start of cooking, min, t = time at point of interest, min, T = temperature of dough, °K, and the other terms have been discussed as part of Equations 3.8 and 3.9. Temperature as a function of time is required to perform the integration necessary to evaluate the apparent viscosity. The above model appears most useful for correlation of viscosity changes which occur during the initial stages of denaturation only. After long time intervals, the model predicts an infinite viscosity which is not practical.

Morgan et al.[15a] have developed a theoretical model which describes the effects of temperature-time history, temperature, shear rate, and moisture content on the apparent viscosity of defatted soy flour dough. Their model uses the Casson rheological equation which defines a yield stress and finite viscosities at high shear rates and is written as

$$\eta = \eta^* \exp(\Delta E_\eta / RT + bM) \left[\sqrt{\frac{\tau_o}{\dot{\gamma}}} + \sqrt{\mu_o} \right]^2$$
$$\left[1 + B(1 - M)^m (1 - \exp(-k_\infty \psi))^m \right] \qquad (3.12B)$$

where b, B, m = constants, τ_o = yield stress, μ_o = limiting viscosity at high shear rates, k_∞ = specific reaction velocity constant, ψ = integral temperature-time history function, given as

$$\psi = \int_0^t \exp(-\Delta E_k / RT) \, dt \qquad (3.12C)$$

Again, to evaluate this model, temperature as a function of time is required. In a computer simulation of a food extruder, performed incrementally, it would be possible to predict the temperature of the dough resulting from the viscous dissipation of the mechanical energy input and heat transfer with an iterative procedure to match temperatures and viscosities which are interdependent.

No single model exists describing apparent viscosity as a function of composition, variability of ingredients, time, temperature, and other extrusion parameters. The power law model has been used to describe the isothermal flow through the metering section of an extrusion screw by Harmann and Harper[6] and Tsao et al.,[22] and for cooking extrusion by Fricke et al.[5] Continued work is required to develop and test the most suitable form of an apparent viscosity model that adequately, yet simply, describes the important changes in viscosity that occur with time during cooking of food doughs. Once a model exists, it will be theoretically possible to simulate the entire cooking extrusion process leading to an increased understanding of the principal variables and their influence on extrusion rate, energy input and finished product characteristics.

C. Other Non-Newtonian Effects

Biopolymeric food materials exhibit a variety of non-Newtonian effects other than the nonlinear tangential relationship between τ and $\dot{\gamma}$ previously discussed. Many of these non-Newtonian effects can be described as viscoelastic which means the fluids behave as a combination of pure elastic behavior, demonstrated by solids and often likened to a spring, and viscous behavior as demonstrated by the dashpot. Various viscoelastic models have been proposed which consist of the spring and dashpot elements placed in series (Maxwell body) or parallel (Kelvin body). The Maxwell body behaves more like a fluid since it would shear indefinitely as long as a constant stress is applied. Conversely, the Kelvin body behaves more like a solid since its strain, γ, remains finite. The characteristics of the members making up these models can be determined by measuring their response to a forcing function having varying frequency.

Viscoelastic behavior is of particular interest when dealing with unsteady state processes where the time response is critical. Food extrusions are normally described as steady state operations of flow through the channels of the extrusion screw, dies, etc., which minimizes the necessity of knowing the viscoelastic properties of food doughs.

Time-dependent shear effects under steady state conditions are another form of non-Newtonian behavior. Fluids can show a decrease in η at constant shear stress and are called thixotropic; those which show an increase in η are called rheopectic. Normally there are limits to the amount of the time-dependent effects because they are due to a structural rearrangement of the fluid. The fluid in an extruder is normally subjected to considerable shear before it enters the screw which again minimizes the necessity to consider time-dependent affects.

In addition to the relationships between $\dot{\gamma}$ and the tangential τ, it has been shown that normal stresses are produced by many fluids. Because normal stresses were first measured by Weissenberg, they are often termed the "Weissenberg effect". The existence of normal stresses in biopolymeric food doughs can be seen at the die of the food extruder as an increase in the diameter of the product immediately after it leaves the die. Many times the product swelling is also due to the flash off of superheated steam in the product, and these two independent phenomena should not be confused.

The existence of normal stresses during the shearing of a fluid can cause an appreciable force, normal to the direction of motion. This normal force tends to separate the rotating discs or plates sometimes used to measure apparent viscosity, leading to erroneous measurements or causing the fluid to climb up the shaft of a rotating member, emptying out the fluid container. These two effects cause measurement problems in the field of rheology.

No known work has been published on the magnitude of the normal stresses developed by food doughs. These stresses have been shown to be large for plastic polymers (Metzner[13]) and are probably not negligible in foods. Understanding the normal stresses occurring during the flow of biopolymers would be useful in describing the swelling of the extrudate at the die, improve the calculations of the strength of extruder barrels, and allow better prediction of pressure drops associated with flow through varying geometrical shapes.

IV. VISCOSITY UNITS

No consistent set of units have been used to report viscosity data in the literature. Since apparent viscosity is the ratio of shear stress to shear rate, appropriate units of viscosity are those which incorporate units of force. It is not uncommon, however, to see viscosity expressed in units containing mass rather than force. The relationship between force and mass is expressed by Newton's second law

$$F = ma/g_c \qquad (3.13)$$

where F = force, m = mass, a = acceleration, and g_c = constant required to keep the units consistent. Depending upon the system of units used, g_c can take on the following forms:

$$g_c = \frac{1 \text{ g cm}}{\text{dyne s}^2}$$

$$g_c = \frac{1 \text{ kg m}}{\text{N s}^2}$$

$$g_c = \frac{32.2 \text{ lb}_m \text{ ft}}{\text{lb}_f \text{ s}^2}$$

In this book, apparent viscosities will always be expressed in units of force, however, the relationship between viscosities expressed in units of force and mass is

Table 1
CONVERSION FACTORS FOR SOME
COMMONLY USED UNITS FOR
APPARENT VISCOSITY

	To convert from	To	Multiply by
η_F	Poise	$\dfrac{dynes-s}{cm^2}$	1
	Poise	$\dfrac{N \cdot s}{m^2}$	0.1
	Cp	poise	0.01
	Cp	$\dfrac{lb_f \cdot s}{ft^2}$	2.09×10^{-5}
η_m	Poise	$\dfrac{g}{cm \cdot s}$	1
	Poise	$\dfrac{kg}{m \cdot s}$	0.1
	Cp	$\dfrac{lb_m}{ft \cdot s}$	6.72×10^{-4}
	Cp	$\dfrac{lb_m}{ft \cdot h}$	2.42

$$\eta_F = \eta_m/g_c \tag{3.14}$$

where η_F = apparent viscosity expressed in units of force, η_m = apparent viscosity expressed in units of mass, and g_c = constant defined above. Table 1 gives some of the conversion factors which make it relatively easy to convert apparent viscosities in one set of units to another set which is more suitable for the user's purposes.

V. VELOCITY PROFILES

The rheological behavior of a fluid profoundly influences the shape of the velocity profile which results as the fluid flows through some type of conduit. The velocity profiles which result in conduits having cross-sections which are circular or a narrow slit are of particular interest in food extrusion and will be described in more detail below.

The force balance on a cylindrical element inside a round tube is shown diagrammatically in Figure 6. Summing forces on the element gives

$$(P + \Delta P)(\pi r^2) - P(\pi r^2) = \tau(2\pi rL) \tag{3.15}$$

showing the difference in the pressure on the ends of the element times its cross-sectional area is equal to the shear stress at the surface of the element times the surface area. Solving for τ yields

$$\tau = \Delta Pr/2L \tag{3.16}$$

and at the tube wall

$$\tau_w = \Delta PR/2L \tag{3.17}$$

Taking the ratio of Equations 3.16 and 3.17 gives

$$\tau/\tau_w = r/R \tag{3.18}$$

The relationship between τ and the shear rate or velocity gradient is given by Equation 3.2 for a Newtonian fluid and can be written as

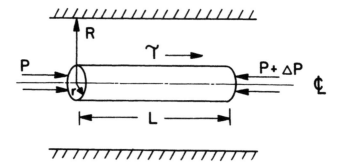

FIGURE 6.   Force balance on a cylindrical element in a circular tube.

$$\frac{dv}{dr} = -\frac{\tau}{\mu} = -\frac{\tau_w r}{\mu R} \qquad (3.19)$$

Separating the variables in Equation 3.19 and performing the indicated integration gives

$$\int_0^v dv = -\frac{\tau_w}{\mu R} \int_R^O r dr \qquad (3.20)$$

or

$$v = \frac{\tau_w}{2R\mu} (R^2 - r^2) \qquad (3.21)$$

Recognizing the maximum velocity will occur at r = 0,

$$v_{max} = \tau_w R/2\mu \qquad (3.22)$$

Defining the average velocity as

$$\bar{v} = \dot{m}/\rho s = 1/s \int v ds$$

where $\bar{v}$ = average velocity in conduit, $\dot{m}$ = mass rate of flow, ϱ = fluid density, s = cross-sectional area of conduit, πR^2, ds = $2\pi r dr$, and v = velocity defined by Equation 3.21, then

$$\bar{v} = \frac{\tau_w}{R^3\mu} \int_O^R (R^2 - r^2)dr = \frac{\tau_w R}{4\mu} \qquad (3.23)$$

or

$$\frac{v}{\bar{v}} = 2[1 - (r/R)^2] \qquad (3.24)$$

Equation 3.24 is the equation for a parabola which is the characteristic velocity profile of a Newtonian fluid in a circular conduit.

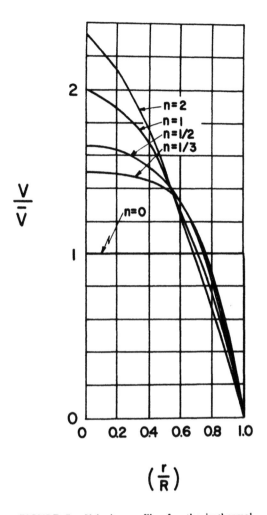

FIGURE 7. Velocity profiles for the isothermal
flow of power law fluids through tubes. (From
McKelvey, J. M., *Polymer Processing*, John Wiley
& Sons, New York, 1962. Copyright © John Wiley
& Sons. Reprinted by permission of John Wiley &
Sons, Inc.)

The above sequence of steps can also be performed using the power law model,
Equation 3.5, to express the relationship between τ and $\dot{\gamma}$. When this is done, the
following results

$$\frac{v}{\bar{v}} = \frac{3n+1}{n+1}\left[1-(r/R)^{\frac{n+1}{n}}\right] \qquad (3.25)$$

indicating how v changes as a function of r for a power law fluid. Note that Equation
3.25 reduces to Equation 3.24 for the case where n = 1. Some velocity profiles for
varying values of n are shown in Figure 7. Observe how the velocity profile becomes
characteristically flatter as n gets smaller to the limiting value of n = 0 where plug
flow results.

A similar procedure to that just described for flow inside tubes can be followed to
determine the velocity profiles for fluids flowing inside long narrow slits. A diagram

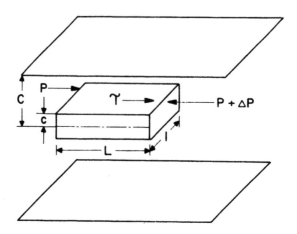

FIGURE 8.   Force balance on a rectangular element between two parallel plates.

of the slit being analyzed is shown in Figure 8. The slit is very wide (infinite) compared to its thickness, 2C, so that the effects of the sides of the slit can be ignored. Again a force balance is written around a fluid element of unit width to determine the relationship τ and ΔP as

$$(P + \Delta P)(2c \cdot 1) - P(2c \cdot 1) = \tau(2L \cdot 1) \tag{3.26}$$

or

$$\tau = \Delta Pc/L \tag{3.27}$$

Rewriting the balance for the shear stress at the wall gives

$$\tau_w = \Delta PC/L \tag{3.28}$$

where 2C is the thickness of the slit. Substituting Newton's law of viscosity into Equation 3.28 and integrating yields

$$v = \tau_w C/2\mu \left[1 - (c/C)^2\right] \tag{3.29}$$

It can be shown that

$$\bar{v} = \tau_w C/3\mu \tag{3.30}$$

so

$$\frac{v}{\bar{v}} = \frac{3}{2}\left[1 - (c/C)^2\right] \tag{3.31}$$

which gives the parabolic velocity distribution across the width of the narrow slit opening. Replacing Newton's law of viscosity with the power law model in the above derivation gives

$$\frac{v}{\bar{v}} = \frac{2n + 1}{2n} \left[1 - (c/C)^{\frac{n + 1}{n}} \right] \qquad (3.32)$$

which reduces to Equation 3.31 for the case where n = 1.

VI. RHEOLOGICAL MEASUREMENTS

The measurement techniques used to determine the rheological properties of food doughs have been reviewed by Clark.[4] A more general description of viscosity and flow measurement techniques and equipment can be found in Van Wazer et al.[23] To be of value in the modeling of food extrusion processes and flow-through dies, rheological measurements must be absolute for the development of constitutive equations rather than the relative measurements that are common to the characterization of many foods for quality control purposes.

A variety of rotational rheometers of the cone and plate or concentric cylinder design are available, but most have limited application to non-Newtonian fluids except for very low shear rates. Above shear rates of 10 s^{-1}, normal forces can become large causing the sample to leave the gap or difficulty in maintaining the precise placement of the rotating elements (Clark[4]). Other problems with the rotational rheometers involve maintaining precise temperature control in the sample. Also, the relatively small sample size required for temperature control makes it difficult to load the measuring cell and control moisture loss when it is heated. For this reason, capillary or slit viscometers have been most universally used for the measurement of food doughs and will be discussed in detail below.

A. Capillary Viscometers

In principle, the measurement of the rheological behavior of fluid materials with a capillary viscometer is very straight forward. A number of assumptions are made concerning the flow processes involved so that the viscosity can be calculated using the standard equations which govern flow through the geometry under consideration. These assumptions are (Van Wazer et al.[23])

1. Flow is steady.
2. No radial or tangential components of velocity exist.
3. Axial velocity is a function of distance from the axis alone.
4. No slippage occurs at the walls.
5. Long tubes are used so that end effects are insignificant.
6. Fluids are incompressible.
7. All external forces are insignificant.
8. Isothermal conditions exist throughout the sample.
9. Viscosity does not change with pressure.

In most cases these assumptions hold or corrections can be made to compensate for the deviation from the assumed condition. In many cases, corrections for end effects are applied so that short capillaries or extrusion dies can be used for measurements.

To determine a dough viscosity, dough at constant temperature is forced through a long, round capillary or slit, and data on the flow rate resulting from different applied pressures is taken. From the pressure measurements, the shear stress can be calculated at the wall using Equation 3.17.

The shear rate can be calculated from the flow recognizing that

$$\bar{v} = Q/\pi R^2 . \tag{3.33}$$

Combining the average velocity calculated above with Equation 3.24 for Newtonian fluids and differentiating with respect to v gives

$$\dot{\gamma} = -dv/dr = 4Qr/\pi R^4 \tag{3.34}$$

or

$$\dot{\gamma}_w = (-dv/dr)_w = 4Q/\pi R^3 = \dot{\gamma}_a \tag{3.35}$$

Recognizing

$$\mu = \tau_w/\dot{\gamma}_w = \frac{\Delta PR/2L}{4Q/\pi R^3} = \frac{\pi \Delta PR^4}{8QL} \tag{3.36}$$

gives the Hagen-Poiseuille equation which can be used to calculate Newtonian viscosity directly from capillary flow data.

In most cases, however, food doughs are non-Newtonian and the apparent viscosity varies with ẏ which varies across the radius of the capillary. Apparent viscosity can be calculated from measurements taken with capillary rheometers when the data are transformed to give τ and ẏ at the same location, which for convenience is the wall. In the non-Newtonian case, τ_w can be calculated from the pressure drop across the capillary directly from Equation 3.17. It is, however, necessary to correct the Newtonian ẏ_w or ẏ_a calculated using Equation 3.35 since for non-Newtonian fluids, ẏ is not linear with respect to r.

Rabinowitsch showed that shear calculated from the average flow rate could be corrected to give the shear rate at the wall of a capillary. Flow rate can be found for any circular cross-section when v as a function of r is known using

$$Q = \int_0^R 2\pi r v dr \tag{3.37}$$

Integrating by parts gives

$$Q = \left|\pi r^2 v\right|_0^R - \pi \int_0^R r^2 \frac{dv}{dr} dr \tag{3.38}$$

Since v = 0 at r = R, because no slip is assumed at the walls, the first term on the right side of Equation 3.38 is zero. The shear stress is directly proportional to r as shown by Equation 3.18, so Equation 3.38 can be transformed to

$$\dot{\gamma}_a = 4Q/\pi R^3 = \frac{4}{\tau_w^3} \int_0^R f(\tau)\tau^2 d\tau \tag{3.39}$$

Differentiating the above with respect to τ_w using the Leibnitz formula gives

$$\dot{\gamma}_w = f(\tau_w) = \frac{3}{4}(\dot{\gamma}_a) + \frac{1}{4}(\dot{\gamma}_a)\left[\frac{d\ln(\dot{\gamma}_a)}{d\ln(\tau_w)}\right] \tag{3.40}$$

Recognizing that (Metzner and Reed[14])

$$\frac{1}{n} = \frac{d\ln\dot{\gamma}}{d\ln\tau_w} = \frac{d\ln(\dot{\gamma}_a)}{d\ln(\tau_w)} \tag{3.41}$$

where n is the flow behavior index in the power law model or the slope of the $\ln \tau_w$ vs $\ln \dot{\gamma}_a$ plot, the shear rate at the wall can be written

$$\dot{\gamma}_w = \frac{3n + 1}{4n} (\dot{\gamma}_a) \tag{3.42}$$

For pseudoplastic materials, where n is constant over a wide range of $\dot{\gamma}_a$ and <1, Equation 3.42 shows that $\dot{\gamma}_w$ will be greater than the shear rate based on the average velocity alone, $\dot{\gamma}_a$. Once both τ_w and γ_w are known, their ratio is the true η which can be correlated with $\dot{\gamma}$ using the power law model.

Similar calculations and corrections can be made to data taken from flow measurements using a narrow slit. For the narrow slit, the average velocity is given by

$$\bar{v} = Q_1 / 2C \cdot 1 \tag{3.43}$$

where Q_1 is the flow rate per unit slit width. Combining the average velocity with Equation 3.31 for the velocity within the slit and differentiating with respect to c gives

$$\dot{\gamma} = -\frac{dv}{dc} = \frac{3Q_1 c}{2C^3 \cdot 1} \tag{3.44}$$

or

$$\dot{\gamma}_w = (-dv/dc)_w = 3Q_1/2C^2 \cdot 1 = 3Q/2C^2 w \tag{3.45}$$

where Q_1 = flow rate per unit slit width, Q/w, Q = flow rate, w = slit width, c = thickness dimension measured from center line of slit, and C = half of slit thickness. For non-Newtonian fluids, the shear rate at walls can be found using

$$\dot{\gamma}_w = \frac{2n + 1}{3n} \left(\frac{3Q}{2C^2 w} \right) \tag{3.46}$$

where n is the flow behavior index in the power law model. Equation 3.46 shows the true shear rate at the wall for non-Newtonian fluids. Taking the ratio of τ_w (Equation 3.28) and $\dot{\gamma}_w$ (Equation 3.46) gives the correct value for η.

B. Piston Capillary Rheometer

Two types of capillary rheometers have been used to measure dough viscosity. One has been a capillary where the force applied to the dough for the flow results from using a compression tester similar to an Instron® and a small piston extruder. The Merz-Colwell extrusion rheometer is an example and shown diagrammatically in Figure 9. Dough is placed in the heated cylinder and brought up to temperature with the heating jacket surrounding the cylinder. Once temperature equilibrium is reached, force is applied to the plunger and the sample is forced through the narrow capillary at the discharge. The flow rate from the device can be varied independently by adjusting the speed at which the plunger travels and the force applied is measured with the compression load cell.

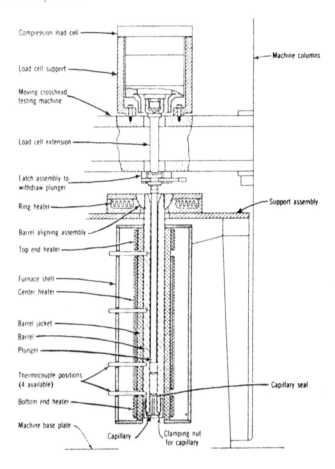

FIGURE 9.   Merz-Colwell extrusion rheometer. (Van Wazer, J. R.,
Lyons, J. M., Kim, K. Y., and Colwell, R. E., *Viscosity and Flow
Measurement,* Interscience, New York, 1963. Copyright © John
Wiley & Sons. Reprinted by permission of John Wiley & Sons.)

The major problems associated with the plunger or piston type of capillary rheometers have been:

1.   Difficulty in loading viscous samples into the narrow and long cylinder. Loading
     usually requires repeated removals of the piston, the addition of a small quantity
     of sample, pushing the sample to the bottom of the chamber with the piston and
     repeating the operation until the chamber is full.
2.   Pressure losses prior to the capillary. If the piston does not move freely, erro-
     neously high pressures will be measured. The added force required to move the
     piston cannot be determined independently making corrections uncertain.
3.   Temperature uniformity within the sample is difficult to achieve because heat
     must be conducted from the walls of the cylinder into the sample. The thermal
     conductivity of the food doughs tends to be low so that an appreciable tempera-
     ture gradient can exist within the sample. To overcome this problem, relatively
     long heating periods are required and when temperatures are high, sample deg-
     radation or nonuniform cooking can occur.
         Heating of the sample can also occur through viscous dissipation as the sample
     is extruded through the capillary. Since more heating occurs at higher $\dot{\gamma}$, temper-

ature gradients can result in the sample during the measurement process which will alter the viscosity and distort the velocity profile.

4.   Pressure drops associated with constriction and expansion losses at the ends of the capillary. These losses result in higher measured total pressures. Corrections for end effects will be discussed later.

Van Wazer et al.[23] list other errors associated with capillary viscometers such as kinetic energy changes, elastic-energy corrections, turbulence, drainage errors, surface tension corrections and wall effects which also may have to be considered. Remsen and Clark[17] and Morgan et al.[15] have successfully used capillary viscometers to measure the apparent viscosity of protein food doughs.

C. Capillary Die

The second type of capillary rheometer uses a food extruder to force a cooked and heated dough through a relatively short die. The advantages of using the extruder for rheological measurements are

1.   No additional equipment is required to make rheological measurements assuming an extruder exists at the laboratory location.
2.   By nature, an extruder will heat, mix, and pressurize the sample behind the capillary or die where the measurements are made. Theoretically, the temperature nonuniformity and sample overheating problems which occur with the plunger type of capillary rheometer should be minimized. The degrees of mixing and temperature uniformity of samples from an extruder are, however, functions of the design and operation of the extruder.

Despite these apparent advantages, use of an extruder has a unique problem along with some of the same problems associated with the plunger-type devices. These are

1.   The operation and flow from an extruder is dependent upon the rheology of the food dough and the back pressure created by the discharge die or capillary. This interdependence makes it difficult to vary the $\dot{\gamma}$ and τs at the die independently. The problem can be partially overcome by varying the die diameter and extruder speed. Clearly, the extruder cannot achieve the independence of operation that the piston-type extruder can produce.
2.   Pressure losses due to end effects on the relatively short die are substantial and must be corrected for the interpretation of the data. Dies with high L/D ratios are not particularly helpful when using an extruder because their pressure drops are also very high at reasonable flow rates so the extruder may not be able to produce very high rates of flow or $\dot{\gamma}$ when long dies are used.

Compensating for the experimental limitations associated with using an extruder to gather rheological data, Harper et al.,[7] Harmann and Harper,[6] Cervone and Harper,[3] Tsao et al.,[22] and Jao et al.[9] have successfully used this technique. Zuilichem et al.[25] used a long slit die to measure the rheology of corn grits.

D. End Corrections

An appreciable part of the pressure drop which occurs across short capillaries can be due to the pressure drops which occur due to the constriction at the entrance of the capillary and expansion losses at the exit. The constriction losses at the entrance can be reduced somewhat if the capillary entrance is tapered rather than an abrupt change in cross-sectional area of the channel for flow. Bagley[1] proposed that the shear stress at the wall of a short capillary could be calculated using

$$\tau_w = \frac{\Delta P}{2\left(\dfrac{L}{R} + \dfrac{L^*}{R}\right)} \qquad (3.47)$$

where τ_w = shear stress at wall, N/m^2, ΔP = pressure drop, N/m^2, L = length of capillary, mm, L* = equivalent length of capillary which would increase ΔP by an amount accounting for end effects, mm, and R = radius of capillary, mm. For relatively short capillaries, L*/R$\sim$5. A similar expression for a slit can be written

$$\tau_w = \frac{\Delta P}{\left(\dfrac{L}{C} + \dfrac{L^*}{C}\right)} \qquad (3.48)$$

where 2C is the width of the narrow slit.

Rogers[19] proposed a calculation procedure for interpretation of capillary flow data which included end effect corrections for the capillary and adjustment of the τ_w calculated assuming Newtonian behavior with the Rabinowitsch correction as suggested by Metzner and Reed[14] and given in Equation 3.42. Use of the procedure requires taking pressure drop and flow data under isothermal conditions using a number of short capillary dies having varying L/R ratios. A plot of log $\dot{\gamma}_a$ vs. log ΔP, where $\dot{\gamma}_a$ = 4Q/πR^3, gives a straight line for data gathered with each die having a different L/R. Next, taking ΔP and L/R points from the first figure for single values of $\dot{\gamma}_a$, a plot of ΔP vs. L/R at different values of $\dot{\gamma}_a$ can be constructed. When the curve for each $\dot{\gamma}_a$ is extrapolated to ΔP = 0, a value for L*/R for each specific value of $\dot{\gamma}_a$ can be determined. A plot of L*/R vs. log $\dot{\gamma}_a$ is linear so that values of L*/R can be determined easily for each $\dot{\gamma}_a$ of interest. Using these values of L*/R, the correct values of τ_w can be calculated with Equation 3.47. Plotting log τ_w vs. log $\dot{\gamma}_a$ from the data yields a line whose slope is n, the flow behavior index. Once n is known, $\dot{\gamma}_w$ can be calculated using Equation 3.42 with η calculated as $\tau_w/\dot{\gamma}_w$. A plot of log η vs. log $\dot{\gamma}_w$ gives a straight line whose slope is n−1 (see Equation 3.8). The calculation procedure is illustrated in the following example.

E. Example 1

Two dies were used with an extruder as a short capillary rheometer. The dimensions of the dies were:

Die	D mm	L mm
1	5.0	18
2	3.0	18

The following data were taken for dough at 120°C.

	Die 1		Die 2	
Reading	Q cm³/30 sec	ΔP atm	Q cm³/30 sec	ΔP atm
1	8.7	8.9	6.8	23.7
2	19.5	12.9	7.3	24.4
3	32.1	16.4	8.0	25.5
4	46.8	19.5	11.4	29.9
5	71.1	23.8	14.2	32.9

Using the power law model (Equation 3.7) to correlate the data, estimate m and n. For the solution:

1. Calculate $\dot{\gamma}_a = 4Q/\pi R^3$.

$$\text{Die 1} \quad \dot{\gamma}_a = \frac{4Q}{\pi R^3} = \frac{4Q \text{ cm}^3}{30 \text{ s}} \left| \frac{1}{\pi} \right| \frac{1}{(0.25)^3 \text{ cm}^3} = 2.716\ Q$$

$$\text{Die 2} \quad \dot{\gamma}_a = \frac{4Q}{\pi R^3} = \frac{4Q \text{cm}^3}{30 \text{ s}} \left| \frac{1}{\pi} \right| \frac{1}{(0.15)^3 \text{ cm}^3} = 12.57\ Q$$

2. Plot on Figure 10 log $\dot{\gamma}_a$ vs. log ΔP.
3. From Figure 10 find ΔP required at $\dot{\gamma}_a = 50$ and 160 s⁻¹ for each die.

$\dot{\gamma}_a$ s⁻¹	ΔP, atm	
	L/R = 7.2	L/R = 12
50	12.6	18.7
160	21.7	32.2

These data are then plotted on Figure 11.

4. From Figure 11, determine L*/R for the two shear rates used by extrapolating the plot to ΔP = 0.

$\dot{\gamma}_a$ s⁻¹	L*/R
50	2.93
160	3.54

5. Plot L*/R vs log $\dot{\gamma}_a$ on Figure 12. From Figure 12 obtain a value for L*/R for each data point.
6. Calculate $\tau_w = \Delta P/[2(L/R + L*/R)]$ for each data point.

$$\tau_w = \frac{\Delta P \text{ atm} \left| \dfrac{1.0132 \times 10^5 \text{ N}}{\text{m}^2 \cdot \text{atm}} \right|}{2\left(\dfrac{L}{R} + \dfrac{L^*}{R}\right)}$$

7. Plot Figure 13, log τ_w vs. log $\dot{\gamma}_a$. Estimate slope from figure which is flow behavior index, n = 0.42.
8. Calculate $\dot{\gamma}_w$.

$$\dot{\gamma}_w = \frac{3n + 1}{4n}\ \dot{\gamma}_a = \frac{3(0.42) + 1}{4(0.42)}\ \dot{\gamma}_a = 1.345\ \dot{\gamma}_a$$

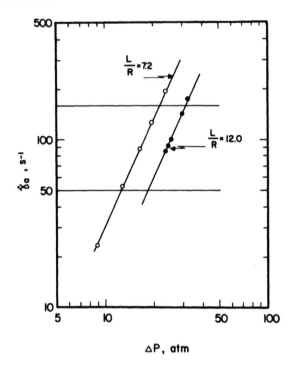

FIGURE 10.   Plot of log $\dot{\gamma}_a$ vs log ΔP.

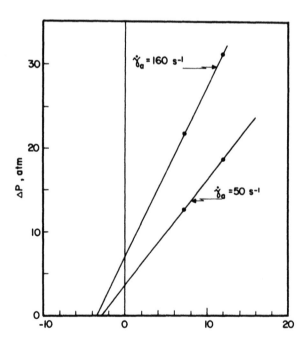

FIGURE 11.   Plot of ΔP vs L/R for $\dot{\gamma}_a$ = 50 and 160 s^{-1}.

9.   Calculate $\eta = \tau_w/\dot{\gamma}_w$.
10.   Plot Figure 14, log η vs. log $\dot{\gamma}$
    From figure slope = -0.58 = n$-$1
    From intercept, m = 10,800 (N$_s^{0.42}$)/m^2

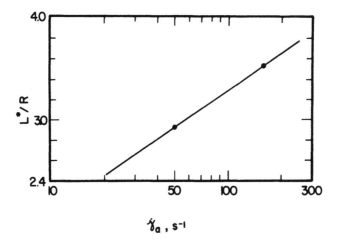

FIGURE 12.   Plot of L*/R vs $\dot{\gamma}_a$.

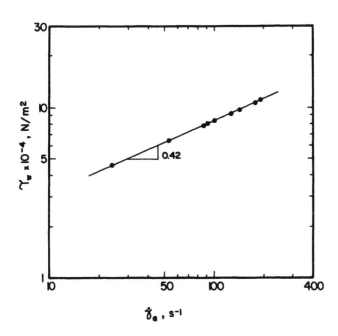

FIGURE 13.   Plot log τ_w vs log $\dot{\gamma}_a$.

11.   Final model for dough at 120°C.

$$\eta = 10.8 \times 10^3 \, \dot{\gamma}^{-0.58}$$

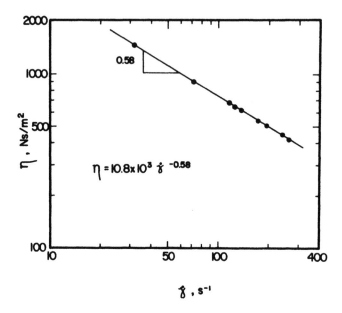

FIGURE 14.   Plot log η vs log $\dot{\gamma}$.

Summary of Calculations

Die	Reading	γ_a s^{-1}	L/R	L*/	$\tau_w \times 10^{-4}$ N/m^2	γ_w s^{-1}	η Ns/m^2
1	1	23.6	7.2	2.56	4.62	31.7	1460
1	2	53.0	7.2	2.96	6.43	71.3	902
1	3	87.2	7.2	3.22	7.97	117	681
1	4	127	7.2	3.42	9.30	171	544
1	5	193	7.2	3.64	11.1	260	427
2	1	85.5	12.0	3.21	7.90	116	681
2	2	91.8	12.0	3.24	8.12	124	655
2	3	101	12.0	3.29	8.45	136	621
2	4	143	12.0	3.48	9.77	192	509
2	5	178	12.0	3.59	10.7	239	458

VII. OTHER VISCOMETERS

A. Cone and Plate

Although the emphasis on rheological measurements have focused on the use of capillaries or slits in this chapter, other types of rheometers exist. Many do not appear to have the capability of handling the viscous food dough for the reasons already discussed. A cone and plate viscometer, known as the Weissenberg Rheogoniometer,® is an example of an instrument which appears to have the capability to make measurements on food doughs. Van Wazer et al.[23] describe the instrument which is very sophisticated. In addition to making the rather simple measurements already described, it is capable of making simultaneous measurements in both tangential and normal directions to the sample cell over a wide range of shear rates using fluids with high viscosities. The temperature of the cell can be precisely controlled and with a hood to prevent moisture loss, would yield a variety of interesting and meaningful measurements to those interested in food dough rheology.

B. Brabender® Visco-Amylo-Graph

The Brabender® Visco-Amylo-graph is a device which has been used extensively in the food field to characterize the consistency of starches and amylase activity. The instrument varies depending upon its intended purpose with some having a bowl with constant rotational speed, while others have a rotational speed variable from 20 to 300 rpm. The measuring head consists of the rotating bowl into which the sample and a stationary bob are placed. A standard configuration has a bowl with vertical pins which intermesh with similar pins on the bob. Readings are arbitrary units called "Brabender Units" which correspond to a viscous drag measurement. To change the range of measurements, an interchangeable torsion-spring cartridge can be easily replaced in the top of the head.

The Amylo-graph can be heated while making measurements. One of the standard tests on starchy materials consists of a heating cycle starting at 29°C at the rate of 1.5°C/min for 44 min, a constant temperature hold for 16 min at 90°C followed by cooling at 1.5°C/min for 30 min to 50°C.

Because of the complex geometry of the measuring head on the Brabender Visco-Amylo-graph®, it is difficult to convert the "Brabender Units" to absolute rheological values which would be useful for modeling a food extruder. Wood and Goff[24] developed a technique which partially overcomes these problems. Using Newtonian and non-Newtonian fluids having similar viscosities, they determined that the effective shear rate at a bowl speed of 75 rpm was between 40 to 50 s^{-1} when the viscosity of the fluid varied between 0 to 30 Ns/m^2. Calibration on the same unit showed 1 Brabender unit = 1.05 cp with the 350 cm·g sensitivity head or 0.75 cp with the 250 cm·g sensitivity head. Because of differences in the bowl, pin, and bob geometries, it would appear desirable to calibrate each instrument if absolute values of viscosity are required. Even with calibration, the instrument has limitations on shear rates and viscosity which can be handled making its use in measuring food doughs for extrusion limited.

C. Torque Rheometer

Torque rheometers have been used to study the rheological behavior of food systems. An example is the Brabender Plasticorder® which consists of a heated chamber that fits over two irregularly shaped rollers. A quantity of food dough which will just fill the sample chamber is charged and heated. The torque required to turn the rolls at different speeds is measured through a torque arm device and the temperature of the dough is measured with a thermocouple.

Blyler and Doane[2] have developed procedures to relate the data from a Brabender Plasticorder® to that taken with a capillary rheometer. Their results show good correlations for polyethylenes and the ranges of η and $\dot{\gamma}$ measured would cover those found with food doughs. By varying the temperature, these authors were also able to predict ΔE_η, the flow activation energy as described in Equation 9. In handling food doughs at high temperatures, precautions will have to be taken to minimize moisture losses in the heated chamber.

D. Extrusion Measurements

Theoretically, it is possible to estimate viscosities from extrusion data where flow, pressure drops, screw speed, and extruder geometry are known. Chapter 4 discusses the effects of these parameters on extrusion flow rate. Using the data collected during extrusion listed above, it is possible to use Equation 4.22 and back calculate viscosity, the only unknown parameter. The method has some shortcomings, however, which are discussed below.

Table 2
CONSTANTS FOR EQUATION 11 EXPRESSING THE VISCOSITY OF FOOD MATERIALS AT EXTRUSION CONDITIONS

Dough material	T (°C)	η^* (Ns*/m²)	n	$\Delta E_\eta/R$ (°K)	K	Viscosimeter	Ref.
1 Cooked cereal dough 80% corn grits 20% oat flour added M = 25—30%	67—100	7.85×10^1	0.51	2500	−7.9^a	Capillary tube	7
2 Pregelatinized corn flour M = 22—35%	90—150	3.6×10^1	0.36	4390	−14^b	Extrusion die	3
3 Soy grits M = 32%	35—60	7.92×10^{-1}	0.34	3670	0	Capillary tube	17

^{a} Times moisture fraction wet basis.
^{b} Times moisture fraction dry basis.

1. The equations for extruder flow were derived for Newtonian flow behavior and food doughs are very non-Newtonian.
2. It is very difficult to characterize $\dot{\gamma}$ in the extrusion screw, making it hard to properly correlate the data.
3. Achieving isothermal conditions in the extruder is hard because of viscous heating of the food so that correlations with temperature are questionable.
4. The effective length of the extrusion screw is questionable, which causes errors in applying the extrusion flow equations.

To circumvent these problems, Rogers[19] showed that extruder input torque measurements could be used with an effective shear rate at the surface of the screw to estimate the rheological behavior of polyethylene. Torque measurements can be converted to τ at the surface of the extrusion screw. The effective shear rate at the surface of the extrusion screw was estimated using the shear rate for a power law fluid at the surface or the inner cylinder of a coaxial rheometer.

VIII. DOUGH RHEOLOGY DATA

Reported values for the viscosity of food doughs used in extrusion processes are relatively limited. All the values reported, to date, indicate that food doughs can be adequately characterized by the power law model and that they all are pseudoplastic with n < 1.

Table 2 summarizes the literature data on food materials at extrusion conditions which have been correlated using Equation 3.11. An examination of the constants indicates that at higher temperatures and moistures, the viscosity of the food doughs are reduced as would be expected. Typical viscosities of these materials were in the range of 10^3 to 10^4 Ns/m². Using an average value of $\Delta E_\eta/R$ for all samples as 3500 cal/g mol, a temperature change from 100°C to 110°C would result in a 22% decrease in the absolute viscosity. This calculation points up the sensitivity of dough viscosity to even relatively small changes in temperature.

Similarly, using an average value of K, (the moisture coefficient) of −11, a 1% change in moisture from 29% to 30% results in a 10% reduction in absolute viscosity. Clearly, small variations in the moisture content of the food dough will greatly affect operation of the extruder and back pressure produced behind the die plate at a certain flow.

Table 3
CONSTANTS FOR POWER LAW MODEL (EQUATION 8) EXPRESSING THE VISCOSITY OF FOOD MATERIALS AT EXTRUSION CONDITIONS

Dough material	M (%)	T (°C)	m (Ns^n/m^2)	n	Viscosimeter	Ref.
1 Corn grits	13	177	2.8×10^4	0.45—0.55	Slit die	25
	13	193	1.7×10^4			
	13	207	0.76×10^4			
2 Full fat soy beans	15—30	+120	3.44×10^3	0.3	Extrusion model	5
3 Moist food product	35	95	2.23×10^2	0.78	Round die	22
4 Pregelatinized corn flour	32	88	1.72×10^4	0.34	Round die	6
5 Sausage emulsion	63	15	4.3×10^2	0.21	Capillary tube	21
6 Semolina flour	30	45	2.0×10^4	0.5	Capillary tube	16
7 Soy grits	22	160	6.71×10^2	0.75	Round die	19
	25	160	2.98×10^2	0.65		
	32	100	2.88×10^4	0.19		
	32	130	2.86×10^4	0.18		
	32	160	1.78×10^4	0.16		
8 Wheat flour	43	33	4.45×10^3	0.35	Rotating concentric cylinders	10

:

More data exists on food doughs correlated to the power law model (Equation 3.8) at isothermal conditions. These data are summarized in Table 3 and again show the strongly pseudoplastic behavior of food doughs. The variability of the consistency index, m, and flow behavior index, n, indicates that the viscosities are very dependent upon the composition of the food product and the temperature at which the data were taken.

REFERENCES

1. **Bagley, E. B.,** End correction in capillary flow of polyethylene, *J. Appl. Phys.,* 28, 624, 1957.
2. **Blyler, L. L., Jr. and Doane, J. H.,** An analysis of Brabender torque rheometer data, *Polym. Eng. Sci.,* 7, 178, 1967.
3. **Cervone, N. W. and Harper, J. M.,** Viscosity of an intermediate moisture dough, *J. Food Proc. Eng.,* 2(1), 83, 1978.
4. **Clark, J. P.,** Dough rheology, *Food Technol. (Chicago),* 32(7), 73, 1978.
5. **Fricke, A. L., Clark, J. P., and Mason, T. F.,** Cooking and drying of fortified cereal foods: extruder design, *Chem. Eng. Prog. Symp. Ser.,* 73(163), 134, 1977.
6. **Harmann, D. V. and Harper, J. M.,** Modeling a forming foods extruder, *J. Food Sci.,* 39, 1099, 1974.
7. **Harper, J. M., Rhodes, T. P., and Wanninger, L. A., Jr.,** Viscosity model for cooked cereal doughs, *Chem. Eng. Prog. Symp. Ser.,* 67(108), 40, 1971.

8. **Holdsworth, S. D.,** Applicability of rheological models to the interpretation of flow and processing behavior of fluid food products, *J. Texture Stud.,* 2, 393, 1971.
9. **Jao, Y. C., Chen, A. H., Lewandowski, C., and Irwin, W. E.,** Engineering analysis of soy dough rheology in extrusion, *J. Food Proc. Eng.,* 2(1), 97, 1978.
10. **Launay, B. and Bure, J.,** Application of a viscometric method to the study of wheat flour doughs, *J. Texture Stud.,* 4, 82, 1973.
11. **Lens, J. W.,** M.S. thesis, Agricultural University, Wageningen, Netherlands, 1975.
12. **McKelvey, J. M.,** *Polymer Processing,* John Wiley & Sons, New York, 1962, 6.
13. **Metzner, A. B.,** Flow behavior of thermoplastics, in *Processing of Thermoplastic Materials,* Bernhardt, E. C., Ed., Robert K. Krieger, Huntington, N.Y., 1969.
14. **Metzner, A. B. and Reed, J. C.,** Flow of non-Newtonian fluids — correlation of the laminar, transition and turbulent flow region, *AIChE J.,* 1, 434, 1955.
15. **Morgan, R. G., Suter, D. A., and Sweat, V. E.,** Design and modeling of a capillary food extruder, *J. Food Proc. Eng.,* 2(1), 65, 1978.
15a. **Morgan, R. G., Suter, D. A., and Sweat, V. E.,** Modeling the effects of temperature-time history, temperature, shear rate and moisture on apparent viscosity of defatted soy flour dough, Paper No. 79-6002, American Society of Agricultural Engineers, St. Joseph, Mi.
16. **Nazarov, N. I., Azarov, B. M., and Chaplin, M. A.,** Capillary viscometry of macaroni dough, *Izv. Vyssh. Uchebn. Zaved. Pishch. Tekhnol.,* 1971(2), 149, 1971.
17. **Remsen, C. H. and Clark, J. P.,** A viscosity model for a cooking dough, *J. Food Proc. Eng.,* 2(1), 39, 1978.
18. **Rossen, J. L. and Miller, R. C.,** Food extrusion, *Food Technol. (Chicago),* 27(8), 46, 1973.
19. **Rogers, M. G.,** Rheological interpretation of Brabender Plasticorder (extruder head) data, *Ind. Eng. Chem. Process Des. Dev.,* 9(1), 49, 1970.
20. **Roller, M. B.,** Characterization of the time-temperature-viscosity behavior of curing β-staged epoxy resins, *Polym. Eng. Sci.,* 15(6), 406, 1975.
21. **Toledo, R., Cabot, J., and Brown, D.,** Relationship between composition, stability and rheological properties of raw comminuted meat batters, *J. Food Sci.,* 42, 726, 1977.
22. **Tsao, T. F., Harper, J. M., and Repholz, K. M.,** The effects of screw geometry on extruder operational characteristics, *AIChE Symp. Ser.,* 74(172), 142, 1978.
23. **Van Wazer, J. R., Lyons, J. W., Kim, K. Y., and Colwell, R. E.,** *Viscosity and Flow Measurements, A Laboratory Handbook of Rheology,* Interscience, New York, 1963.
24. **Wood, F. W. and Goff, T. C.,** Determination of the effective shear rate in the Brabender Viscograph and in other systems of complex geometry, *Staerke,* 25(3), 89, 1973.
25. **Van Zuilichem, D. J., Buisman, G., and Stolp, W.,** Shear behavior of extruded maize, presented at the 4th International Congress of Food Science and Technology, International Union of Food Science and Technology, Madrid, September 23 to 27, 1974.

Chapter 4

EXTRUSION MODELS

I. INTRODUCTION

Food extrusion has developed rapidly during the past 40 years with applications being continually expanded to new areas of food processing. Through these developments, the food engineer and technologist are faced with many extrusion problems such as:

1. Increasing the productive output of an extrusion system
2. Making process modifications to compensate for formulation changes or ingredient variability
3. Explaining why an extruder behaves in a particular manner
4. Designing an optimum extruder or extruder system
5. Scaling-up an extrusion process developed in the laboratory
6. Specifying the control parameters for an extruder
7. Maximizing the useful information gathered through costly experimental runs

Answers to many of these questions to date have been primarily educated guesses rather than well formulated responses based on an analysis of the system using a theoretical basis.

Food extrusion has been primarily an art and only within the last ten years has any serious effort been made to apply rigorous principles to the complicated process. This statement is made not to condemn the art, for it has been the trial-and-error approach, coupled with a keen sense of observation, which has lead to the many successful applications of food extruders. Because of the complexity and variability of food ingredients, reducing food extrusion to a simple mathematical equation is highly unlikely in the foreseeable future. It is this author's contention, however, that the application of basic engineering and scientific principles, as far as possible, can lead to an increased understanding of the complex process and greatly enhance the effectiveness of the technologist in applying the food processing tool called extrusion.

Fricke et al.[13] compared the current status of the development of food extrusion to the parallel (but earlier) development of the extrusion of thermal plastics. In the plastics industry, the screw extruder has seen wide acceptance since 1940 when its initial applications were made on the basis of trial-and-error approaches and experimental observations. Beginning in the late 1940s, a variety of extrusion models were developed which described the fundamentals of the fluid and heat transport in the extrusion of thermal plastics. These efforts have continued and developed to the point where accurate simulations of the extruder can be made. These simulations have been used to scale-up plastics extrusion operations, reduce experimental time, and optimize the design and operation of extrusion systems. In the food extrusion area, the application of models to actual processes began about 15 years ago in industrial food laboratories and is continuing today with some universities beginning to take an active part.

The extensive theoretical developments made in the area of plastics extrusion is an excellent place to begin the development of a model of a food extruder. However, it is necessary to recognize the similarities as well as the vast differences in materials which are extruded in these two instances. Plastics are relatively homogeneous polymeric materials that can be characterized chemically and physically. In most cases, only a single type of plastic is extruded at one time. On the other hand, food extrusion uses an infinite variety of food ingredients as feed materials. Often the ingredients are biopolymers of starch or protein, but their exact character depends upon their source, age, prior treatment, etc. In plastics extrusion, melting or change of phase of the polymer is the major change which occurs in the extrusion process. In food extrusion, the ingredients and water react in a complex cooking process which causes extensive alterations in the chemical and physical nature of the extruded materials. Consequently, food extrusion is a much more complex process than plastics extrusion. The extensive models describing the melting that occurs in plastics extruders are largely irrelevant to food extruders.

In this chapter the pertinent aspects of extrusion models developed for plastics extrusion will be covered. The relationship and applications of these models to food extrusion will be emphasized and illustrated. Although rigorous application of models to food extrusion is in the future and perhaps even impossible, a thorough understanding of the basic principles of extrusion is essential to augment the art which has dominated the developments in the field.

II. FLOW

A. Isothermal Newtonian Flow
1. Simplified Theory

Modeling of the food extruder has principally focused on the metering section of the screw. This has been a logical choice because it is the metering section that controls the extrusion rate, accounts for the majority of the power consumption, and causes the uniform pressure which occurs behind the die. The metering section of the screw, also, is more amenable to modeling because the geometry of this section is normally fixed and characteristics of food materials are more uniform than they are in either the feed or transition sections of the screw. Modeling food extruders borrow heavily from the developments in the plastics extrusion[6-8,31] for the metering section of the screw. Summaries of this work are also available.[23,25,32]

To solve the basic flow equations, the following assumptions are made:

1.   Flow is laminar.
2.   Flow is steady.
3.   Flow is fully developed.
4.   Barrel is rotating and the screw is stationary.
5.   Channel is "peeled off" the screw and laid flat.
6.   Slip does not occur at the walls.
7.   Fluid is incompressible.
8.   Gravity forces are negligible.
9.   Inertial forces are negligible.
10.   Material being extruded is Newtonian.

Some of these assumptions are suitable for food extrusion while others should be questioned. Food doughs are highly viscous (see Chapter 3) and the screw turns relatively slowly, resulting in Reynolds numbers of less than 10^{-3} which is well within the laminar region. Flow in food extrusion is held steady and velocity profiles remain constant with time and location. If the extrusion model is restricted to the metering section, fully developed flow occurs.

Keeping the screw stationary and rotating the barrel is much simpler for modeling purposes, for the frame of reference is stationary with respect to the screw. Assuming

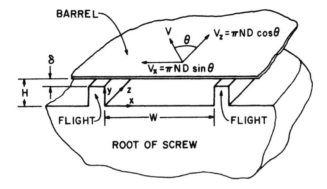

FIGURE 1. Geometry of a shallow channel. (From Tadmor, Z. and Klein, I., *Engineering Principles of Plasticating Extrusion,* Van Nostrand Reinhold, New York, 1970, 357. With permission. © 1970 Society of Plastics Engineers, Inc.)

that the barrel rotates and the screw is stationary is identical to the actual, but reverse, situation, except for the radial pressure distribution resulting from the centripetal forces. The centripetal forces are small because the screw turns slowly.

If the channel of the screw is shallow (H/D is small) it can be unwound from the screw with relatively little distortion and the resulting section laid flat. When this is done, the barrel becomes an infinite plate sliding across the channel at the helix angle θ (as shown in Figure 1).

To assume that no slip occurs at the walls is questionable under a number of food extrusion applications. Grooves in the barel walls reduce slip but complicate modeling. Incorporating slip into the model is difficult and therefore not included.

Cereal doughs are nearly incompressible as determined by Lancaster[21] and Harmann and Harper.[16] Gravity forces in the metering section of an extruder are negligible compared to normal operating pressures of 35 bar (3.5 MPa). Inertial forces are also very small compared to viscous forces.

Since most food doughs do not exhibit Newtonian flow behavior, applying theoretical developments based on Newtonian flow behavior must be done with care. Food doughs are pseudoplastic which will extensively[9,14] alter velocity profiles within the flights. These effects will be discussed later in the book after the theory for Newtonian fluids is developed.

The Navier-Stokes equations for incompressible flow apply and serve as the basis for the analysis of the metering section. The equations of motion in rectangular coordinates in terms of velocity gradients for a Newtonian fluid with constant density and viscosity are

$$
x: \rho\left(\frac{\partial v_x}{\partial t} + v_x \frac{\partial v_x}{\partial x} + v_y \frac{\partial v_x}{\partial y} + v_z \frac{\partial v_x}{\partial z}\right) = -\frac{\partial P}{\partial x}
$$

$$
+ \mu\left(\frac{\partial^2 v_x}{\partial x^2} + \frac{\partial^2 v_x}{\partial y^2} + \frac{\partial^2 v_x}{\partial z^2}\right) + \rho g_x
$$

(4.1)

$$
y: \rho\left(\frac{\partial v_y}{\partial t} + v_x \frac{\partial v_y}{\partial x} + v_y \frac{\partial v_y}{\partial y} + v_z \frac{\partial v_y}{\partial z}\right) = -\frac{\partial P}{\partial y}
$$

$$
+ \mu\left(\frac{\partial^2 v_y}{\partial x^2} + \frac{\partial^2 v_y}{\partial y^2} + \frac{\partial^2 v_y}{\partial z^2}\right) + \rho g_y
$$

(4.2)

$$\text{z: } \rho\left(\frac{\partial v_z}{\partial t} + v_x\frac{\partial v_z}{\partial x} + v_y\frac{\partial v_z}{\partial y} + v_z\frac{\partial v_z}{\partial z}\right) = -\frac{\partial P}{\partial z}$$

$$+ \mu\left(\frac{\partial^2 v_z}{\partial x^2} + \frac{\partial^2 v_z}{\partial y^2} + \frac{\partial^2 v_z}{\partial z^2}\right) + \rho g_z \tag{4.3}$$

For fully developed flow in the z direction, partial deviations with respect to z and time are zero. Because gravity and inertial forces are neglected, these terms may also be dropped giving:

$$\text{x: } \frac{\partial P}{\partial x} = \mu\left(\frac{\partial^2 v_x}{\partial x^2} + \frac{\partial^2 v_x}{\partial y^2}\right) \tag{4.4}$$

$$\text{y: } \frac{\partial P}{\partial y} = \mu\left(\frac{\partial^2 v_y}{\partial x^2} + \frac{\partial^2 v_y}{\partial y^2}\right) \tag{4.5}$$

$$\text{z: } \frac{\partial P}{\partial z} = \mu\left(\frac{\partial^2 v_z}{\partial x^2} + \frac{\partial^2 v_z}{\partial y^2}\right) \tag{4.6}$$

where P = pressure, μ = viscosity, v = velocity, and x,y,z = direction. The continuity equation reduces to

$$\frac{\partial v_x}{\partial x} + \frac{\partial v_y}{\partial y} = 0 \tag{4.7}$$

because $\partial v_z/\partial z$ is zero since flow in the z direction is fully developed.

To determine extrusion output, the flow in the z direction need only be considered. Equation 4.6 was solved by Rowell and Finlayson[26,27] using the boundary conditions $v_z(x,O) = O$, $v_z(x,H) = V_z$, $v_z(O,y) = O$, and $v_z(W,y) = O$, which indicates no slip at the channel boundaries. Once v_z is found, it can be used to estimate volumetric extruder output

$$Q = p \int_O^H \int_O^W v_z dy dx \tag{4.8}$$

where Q = volumetric extruder output and p = number of channels in parallel. This equation was integrated by Squires[29] to give

$$Q = p\frac{V_z WH}{2} F_d + p\frac{WH^3}{12\mu}\left(\frac{\partial P}{\partial z}\right) F_p \tag{4.9}$$

where $V_z = \pi ND\cos\theta$, $W = \pi D\tan(\theta/p)\cos\theta - e$, and F_d = drag flow shape factor, F_p = pressure flow shape factor. The first term in Equation 4.9, called drag flow, is the flow resulting from viscous drag and is proportional to N. The second term is proportional to the pressure gradient down the screw channel and is called pressure flow. Normally, pressure flow is in an opposite direction to the drag flow since the pressure is highest at the discharge of the extruder, causing a negative pressure gradient, $\partial P/\partial z$. Simply, Equation 4.9 is often written as

$$Q = Q_d + Q_p \tag{4.10}$$

where Q_d = drag flow and Q_p = pressure flow. In many food extrusion applications, the drag flow is much larger than the pressure flow. In these cases, Q will increase linearly with N. The ratio of pressure flow to drag flow is an important parameter in extrusion and is defined as

$$a = -\frac{Q_p}{Q_d} = \frac{H^2}{6V_z\mu}\frac{\partial P}{\partial z}\frac{F_p}{F_d} \tag{4.11}$$

The channel shape correction factors in Equation 4.9 account for the increasing influence that the walls of the channel, created by the flights, exert as H/W becomes large. These factors are given as

$$F_d = \frac{16W}{\pi^3 H} \sum_{i=1,3,5\ldots}^{\infty} \frac{1}{i^3}\tanh\left(\frac{i\pi H}{2W}\right) \tag{4.12}$$

$$F_p = 1 - \frac{192H}{\pi^5 W} \sum_{i=1,3,5\ldots}^{\infty} \frac{1}{i^5}\tanh\left(\frac{i\pi W}{2H}\right) \tag{4.13}$$

These shape factors are plotted in Figure 2 and are equal to 1.0 when H/W = O.

2. Velocity Profiles

To obtain some insights into the velocity profiles which exist in the screw channel, it is appropriate to return to Equation 4.6; assume $\partial v_z^2/\partial x^2$ is zero, which is the case when H/W $\sim$ O, and apply the boundary conditions $v_z(x,O) = O$ and $v_z(x,H) = V_z$ indicating no slip at the root of the channel or at the barrel.

Integrating twice gives

$$v_z = \frac{V_z y}{H} + \frac{(y^2 - Hy)}{2\mu}\left(\frac{\partial P}{\partial z}\right) \tag{4.14}$$

which can be rewritten as

$$v_z = V_z\left[(1 - 3a)(y/H) + 3a(y/H)^2\right] \tag{4.15}$$

where a equals the negative ratio of Q_p/Q_d as defined by Equation 4.11.

Figure 3 shows the velocity profile which exists within the channel in the z, or down channel direction, when the components of drag flow and pressure flow are superimposed. For a Newtonian fluid, the velocity distribution within the channel due to drag is linear, going from 0 at the root of the screw to $V_z = \pi ND\cos\theta$ at the top of the flights. The pressure flow, shown for a negative pressure gradient, produces a characteristic parabolic velocity distribution. The combined flows show a net negative flow at the root of the screw and a positive flow toward the barrel surface. As a result, a circulation is created within the flight with some plane existing where $v_z = O$. At closed discharge, the net flow from the extruder is zero and the drag and pressure flow terms have equal magnitude but opposite signs.

A similar analysis can be made for the cross channel velocity profile. In this case, Equation 4.4 is used with the assumption $\partial v_x^2/\partial x^2$ is small. Applying the boundary conditions

$$v_x(z,O) = 0 \text{ and } v_x(z,H) = -V_x = -\pi DN\sin\theta \tag{4.16}$$

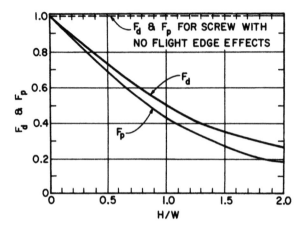

FIGURE 2.   The drag flow (F_d) and the pressure flow (F_p) shape correction factors. (From Squires, P. H., *SPE J.*, 14(5), 24, 1958. With permission.)

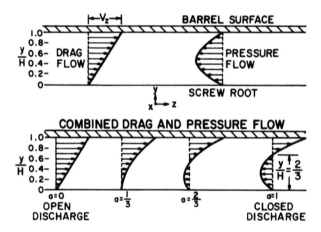

FIGURE 3.   Velocity profile in the down channel z direction showing the superimposed components of drag and pressure flow. (From Paton, J. B., Squires, P. H., Darnell, W. H., Cash, F. M., and Carley, J. F., *Processing of Thermoplastic Materials,* Bernhardt, E. C., Ed., Robert E. Kreiger, Huntington, N.Y., 1974, 178. With permission. © 1970 Society of Plastics Engineers, Inc.)

and integrating twice gives

$$v_x = -\frac{V_x y}{H} - \frac{1}{2\mu} \frac{\partial P}{\partial x} (y^2 - yH) \qquad (4.17)$$

Since no net flow exists in the x direction,

$$\int_0^H v_x dy = 0 \qquad (4.18)$$

because of the existance of flights on the screw. Substituting Equation 4.17 into Equation 4.18 and integrating gives

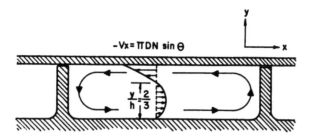

$$-V_x = \pi DN \sin \theta$$

FIGURE 4. Velocity profile in the cross channel x direction. (From Paton, J. B., Squires, P. H., Darnell, W. H., Cash, F. M., and Carley, J. F., *Processing of Thermoplastic Materials*, Bernhardt, E. C., Ed., Robert E. Kreiger, Huntington, N.Y., 1974, 179. With permission. © 1970 Society of Plastics Engineers, Inc.)

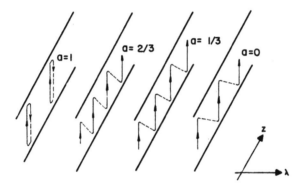

FIGURE 5. The path of a particle in the screw channel as a function of $a = -Q_p/Q_d$. (From McKelvey, J. M., *Polymer Processing*, John Wiley & Sons, New York, 1962, 244. Copyright © 1962 John Wiley & Sons. Reprinted by permission of John Wiley & Sons.)

$$\frac{\partial P}{\partial x} = -6\mu \frac{V_x}{H^2} \tag{4.19}$$

Further substitution of the above into Equation 4.17 gives the velocity profile of the x direction

$$v_x = \frac{y}{H}\left(2 - \frac{3y}{H}\right)V_x \tag{4.20}$$

Figure 4 shows the velocity profile existing within the channel in the x, or cross channel direction. As shown by Equation 4.20, the cross channel velocity is not affected by the down channel pressure gradient and so is independent of the pressure flow. The velocities in the x and z directions must be combined vectorially to obtain the flow in the longitudinal or λ direction. Maximum discharge for an extruder occurs when Q_p = O and zero discharge results when $Q_d = -Q_p$. In any of these flow situations, there is never flow in the channel in the −λ direction. The actual flow of a particle in the channel is a closed helical path as shown in Figure 5. The solid line in this figure represents the path of a particle near the top of the flight, it moves to the bottom of the channel where its path is represented by the dashed line. The direction of the dashed line is a function of a. The particle crosses the channel near the root of the screw and emerges at the surface of the channel once it reaches a flight. Again, notice that in no case is there a time when the particle is moving in the $-\lambda$ direction. It should also be realized that the presence of the flights greatly distorts the velocity profiles in their proximity, requiring the addition of F_d and F_p in the flow equation to correct for this distortion.

3. Corrected Flow Equation

Equation 4.9 can be rewritten in terms of D, e, and L, which are often more convenient geometric parameters, and the rotational speed of the screw, N. Using the definition of V_z and W given under Equation 4.9 and recognizing that

$$L = Z \sin \theta \qquad (4.21)$$

gives

$$Q = G_1 N F_{dt} + \frac{G_2}{\mu} F_{pt} \left(\frac{P_1 - P_2}{L} \right) \qquad (4.22)$$

where

$$G_1 = \frac{\pi^2}{2} D^2 H \left(1 - \frac{ep}{\pi D \sin \theta} \right) \sin \theta \cos \theta \qquad (4.23)$$

$$G_2 = \frac{\pi}{12} DH^3 \left(1 - \frac{ep}{\pi D \sin \theta} \right) \sin^2 \theta \qquad (4.24)$$

$$F_{dt} = F_d F_{de} F_{dc} \qquad (4.25)$$

$$F_{pt} = F_p F_{pe} F_{pc} \qquad (4.26)$$

F_d = drag flow shape factor (Figure 2), F_p = pressure flow shape factor (Figure 2), F_{de} = end correction factor for drag flow (Figure 6), F_{pe} = end correction factor for pressure flow (Figure 7), F_{dc} = curvature correction factor for drag flow (Figure 8), F_{pc} = curvature correction factor for pressure flow (Figure 9), P_1 = pressure at beginning of metering section, and P_2 = pressure at discharge of metering section. The volumetric flow rate, Q, can be converted to a mass flow rate, $\dot{m}$, by multiplying by the dough density, ϱ.

The correction factors, F_{dt} and F_{pt}, account for the deviations from the flat parallel plate model used to derive Equation 4.22 and the channel of a real screw which has sides, oblique ends and is curved and so cannot be laid perfectly flat. The total correction factor is the product of three individual corrections of which F_d and F_p have already been discussed. Booy[2,3] developed the correction factors to account for the oblique ends of the channel (F_{de} and F_{pe}) shown in Figures 6 and 7 and for the channel curvature (F_{dc}, F_{pc}) shown in Figures 8 and 9.

4. Extruder Characteristics

The operating characteristics of a food extruder are determined by coupling the extruders flow with that of the die at the extruder discharge. Once the food product leaves the extrusion screw, the pressure build-up at the tip causes the viscous dough to flow through the die. Common die cross sections are circles, slits, or annuli. For a Newtonian fluid, the Hagen-Poiseuille equation can be applied. Neglecting entrance and exit effects, it can be simply written as

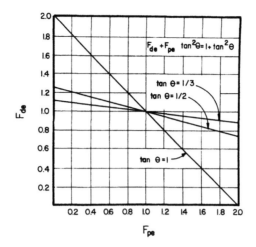

FIGURE 6. Drag flow end correction factor. (From Booy, M. L., *Polym. Eng. Sci.*, 7(1), 5, 1967. With permission. © 1967 Society of Plastics Engineers, Inc.)

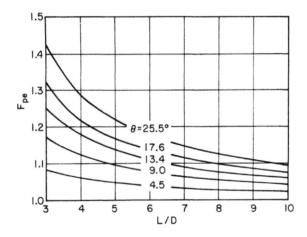

FIGURE 7. Pressure flow end correction factor. (From Booy, M. L., *Polym. Eng. Sci.*, 7(1), 5, 1967. With permission. © 1967 Society of Plastics Engineers, Inc.)

$$Q = K \frac{\Delta P}{\mu_d} \qquad (4.27)$$

where K = geometric constant depending on type of die opening, ΔP = pressure drop across die, and μ_d = viscosity of dough at die. The viscosity of the dough at the die may be different than that in the screw because of differences in temperature and shear rate in these two locations.

Equations 4.22 and 4.27 can be combined to give

$$Q = \frac{G_1 N F_{dt}}{1 + \left(\dfrac{\mu_d}{\mu}\right)\left(\dfrac{G_2 F_{pt}}{LK}\right)} \qquad (4.28)$$

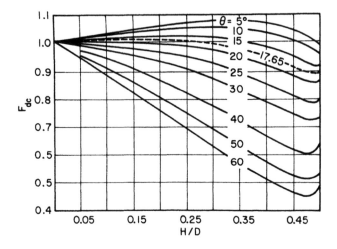

FIGURE 8. Drag flow curvature factor. (From Booy, M. L., *SPE Trans.*, 3(3), 176, 1963. With permission. © 1963 Society of Plastics Engineers, Inc.)

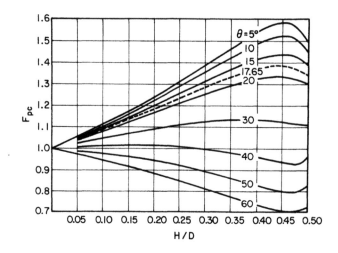

FIGURE 9. Pressure flow curvature factor. (From Booy, M. L., *SPE Trans.*, 3(3), 176, 1963. With permission. © 1963 Society of Plastics Engineers, Inc.)

and

$$\Delta P = \frac{G_1 N F_{dt}}{\left(\dfrac{K}{\mu_d}\right) + \left(\dfrac{G_2 F_{dt}}{\mu L}\right)} \tag{4.29}$$

These equations give the flow resulting from the pressure drop across a specific die with an extruder operating at steady flow conditions when the food material is Newtonian. Graphically, Q vs ΔP relationships are shown in Figure 10. The flow characteristics for two screws, one with a deep channel and the other with a shallow channel, are shown. The output of the deep channel screw with equal N, and D, and ΔP = O is substantially larger than the output of the shallow channel screw because the drag flow is proportional to H. The deep channeled screw output is, however, highly de-

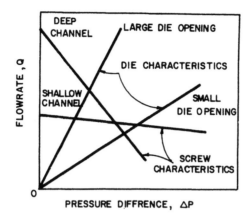

FIGURE 10. Flow as a function of ΔP for varying screw and die characteristics.

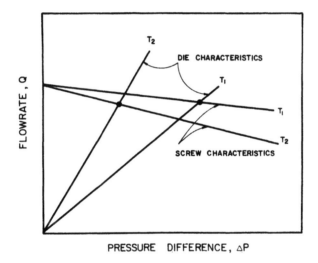

FIGURE 11. Flow as a function of ΔP at to different extrusion temperatures.

pendent upon the pressure flow, and the output decreases rapidly as back pressure increases. The shallow channel screw is relatively unaffected by high back pressure and its output is nearly constant with increasing ΔP.

The flow through a large and small die is also plotted in Figure 10 and is also a function of ΔP. The point where the die characteristic curve crosses the screw characteristic curve represents the operating point of the extruder and is described by Equations 4.28 and 4.29. Notice how going from a large to a small die may actually result in the shallow flighted screw having a greater output than the deep flighted screw.

Figure 11 shows the operation of an extruder and flow through a die at two different temperatures. The viscosity of food doughs are quite sensitive to small changes in temperature with lower viscosities occurring with higher temperatures. Holding all extrusion variables constant except temperature will only affect pressure flow which will increase because it is inversely proportional to μ. The increase in pressure flow causes a net decrease in Q. At the higher temperature, the flow through the die also increases for a given ΔP so the operating point is shifted to the left of the original operating point. It is interesting to note that despite a significant change in temperature the extruder output, Q, is relatively unaffected in this example.

5. Leakage Flow

A clearance, δ, is required between the land of the screw and the barrel to avoid metal-to-metal contact when the screw rotates. Differential thermal expansion of the barrel and screw may cause the clearance to change from that measured when the extruder is cold to that which actually occurs during operation.

The magnitude of the flight clearance effects leakage over the flights, the heat transfer coefficient at the surface of the barrel, and the heat generated by viscous dissipation of the mechanical energy input. The cause of the leakage flow is the cross channel pressure gradient which occurs due to the circulation of the dough within the channel. As the screw wears with time, δ and leakage flow increase and effectively reduce H which also lessens the drag flow.

Further modifications to Equation 4.22 have been made to account for flight clearance. Tadmor and Klein[32] have shown that the effect of clearance can be included as

$$Q = G_1 NF_{dt}\left(1 - \frac{\delta}{H}\right) + \frac{G_2}{\mu}\, F_{pt}\left(\frac{P_1 - P_2}{L}\right)(1 + f) \qquad (4.30)$$

where δ = clearance between flight and barrel,

$$f = G_3\frac{\mu}{\mu_\delta} + G_4\left[\left(\frac{G_5\mu N}{P_1 - P_2} + G_6\right)\Big/\left(1 + \frac{G_7\mu_\delta}{\mu}\right)\right] \qquad (4.31)$$

$$G_3 = (\delta/H)^3\,\frac{e}{W} \qquad (4.32)$$

$$G_4 = 1 + \frac{e}{W} \qquad (4.33)$$

$$G_5 = \frac{-6L\pi D\,(H - \delta)}{H^3\,\tan\theta} \qquad (4.34)$$

$$G_6 = \frac{1 + \dfrac{e}{W}}{\tan^2\theta} \qquad (4.35)$$

$$G_7 = \left(\frac{H}{\delta}\right)^3\,\frac{e}{W} \qquad (4.36)$$

and μ_δ = viscosity in flight clearance.

The drag flow component in Equation 4.30 has the term $(1 - \delta/H)$ as a multiplier which shows the effective reduction in the drag flow as H is reduced by increasing δ. It should be noted that f can change signs depending upon whether the pressure drop is positive or negative down the length of the extrusion screw. Pressure flow increases as δ^3 so that as δ becomes large, it can have a substantial impact on the net flow rate from the extruder.

6. Flow in Barrel Grooves

It is common practice to use grooved barrels on food extruders to reduce the slip at the wall and improve pumping performance. The presence of the grooves does reduce slippage but provides an additional area for leakage flow.

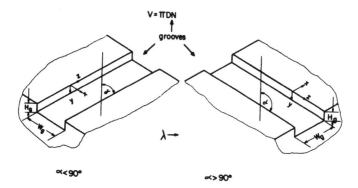

FIGURE 12.   Measurement of the helix angle α for grooves in the barrel relative to the screw tip velocity vector.

To analyze the flow through the grooves in the barrel, assumptions similar to those required to develop the equation for flow in the channel of the screw are required. Only the point of reference need be changed so that the flattened barrel with the grooves is held stationary, and a surface representing the top of the screw is slid across the stationary barrel surface.

The definition of the relative angle between the two sliding surfaces becomes crucial to the analysis. Figure 12 shows the measurement of the angle α and the screw tip velocity vector. Modifying Equation 4.9 to use the dimensions of the grooves gives

$$Q_g = j \left[\frac{\pi D N W_g H_g \cos \alpha}{2} F_d + \frac{W_g H_g^3 \sin \alpha}{12 \mu_g} \left(\frac{P_1 - P_2}{L} \right) F_p \right]$$

(4.37)

where j = number of grooves in parallel, D = diameter of barrel, N = rotation speed of screw, W_g = width of groove, H_g = height of groove, μ_g = viscosity of dough in groove, L = length of barrel, P_2 = pressure at die, P_1 = pressure at feed, α = helix angle of groove defined by Figure 12, and F_d, F_p = wall correction factor given in Figure 2.

To obtain the volumetric extrusion rate, it is necessary to add Q_g to Equation 4.22 giving

$$Q = G_1 N F_{dt} + \frac{G_2}{\mu} F_{pt} \left(\frac{P_1 - P_2}{L} \right) + Q_g$$

(4.38)

when leakage flow over the lands is not important. When leakage flow over the lands of the screw is significant, Q_g should be added to Equation 4.30 to obtain the extrusion output when grooves are present.

Notice that the expression for Q_g consists of a drag term proportional to N and a pressure flow term proportional to ΔP. Flow in the grooves will be large when H_g is large.

The flow direction of dough through the grooves is related to the helix angle of the groove, α, and the pressure drop. If $\alpha < 90°$ and the pressure drop is positive, the Q_g is positive and flow at the extruder discharge is enhanced by the presence of the groove. Since ΔP is rarely positive, the pressure flow term in Equation 4.37 results in a negative flow which is seen as a reduction in extruder output.

When $\alpha = 90°$, there is no drag flow in the grooves and the groove acts only as a channel for flow due to the pressure difference which exists between the feed of the extruder and the die.

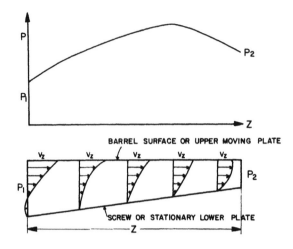

FIGURE 13. Pressure and velocity profile in tapered channel. (From Tadmor, Z. and Klein, I., *Engineering Principles of Plasticating Extrusion,* Van Nostrand Reinhold, New York, 1970, 237. With permission. © 1970 Society of Plastics Engineers, Inc.)

7. Tapered Channels

Tapered channels have been used in food extrusion where the flight height gradually decreases from the feed to the discharge of the screw. Equation 4.9 can be applied to the conditions which exist at any point in the down channel direction. Assuming a linear relationship between H and z such that

$$H = H_1 - \frac{H_1 - H_2}{Z} z \qquad (4.39)$$

where H = height of channel, Z = length of channel, z = distance down channel, and subscripts 1 = start of channel and 2 = end of channel, Equation 4.9 can be integrated to give a relationship for P and z. In order to achieve the same flow through each channel cross-section, a maximum in pressure may be reached at some point down the channel. The existence of the maximum pressure depends upon the initial values of P_1 and P_2. Figure 13 illustrates how the pressure and velocity profiles can vary with the channel length. At the point where the pressure is at maximum, pure drag flow will exist within the channel.

The integrated flow equation for tapered channels can be written as

$$Q = \rho \frac{WH_2 V_z}{2}\left(\frac{2}{1 + \dfrac{H_2}{H_1}}\right) - \rho \frac{WH_2{}^2}{12\mu}\left(\frac{P_2 - P_1}{L}\right)\left[\frac{2}{\dfrac{H_2}{H_1}\left(1 + \dfrac{H_2}{H_1}\right)}\right] \qquad (4.40)$$

where the corrections for the flights have been assumed equal to 1.0. Use of correction factors, F_{dt} and F_{pt}, based on the average flight height to account for the presence of flights, channel curvature and oblique end effects would be appropriate when Equation 4.40 is applied. W and V_z are defined below Equation 4.9.

B. Isothermal Non-Newtonian Flow
1. Application of Simplified Theory

The fact that food doughs are highly pseudoplastic leads to a complication in the solution of the equations of motion. In the case of Newtonian flow, the solution results in two independent terms, Q_d and Q_p, which greatly assists in the interpretation of the flow phenomena occurring in the channel of the screw. When non-Newtonian doughs are used, drag and pressure flow terms become dependent and calculation of flow rates using Equation 4.22 can lead to errors. When pressure gradients are positive, Q is underestimated; when they are negative, Q is overestimated. The reason for this change is because the velocity profiles in the screw are greatly changed when drag and pressure flows are combined depending upon the direction of the pressure flow. When the velocity profiles change, $\dot{\gamma}$ changes which affects η.

Despite these difficulties in applying the simplified models for flow based on Newtonian viscosity, they can be used as a first approximation with reasonable accuracy where η is substituted directly for μ in Equation 4.22. In these cases η is evaluated at an effective shear rate,

$$\dot{\gamma}_H = \frac{\pi DN}{H} \qquad (4.41)$$

More sophisticated techniques can be used to evaluate an effective shear rate but these are probably not warranted since they will not correct the basic problem of drag and pressure flows not being independent with non-Newtonian fluids. The major salvaging factor is that in most food extrusion, the pressure flow is small so that more error can be tolerated in the evaluation of Q_p without it greatly affecting the total flow, Q.

2. Example 1

1. Calculate the volumetric flow rate for the metering zone of a food extruder when a cooked corn dough (Harper, Rhodes, and Wanninger[19]) at 80°C is being used. The pertinent dimensions of the metering section are p = 1, single flighted screw, D = 6.35 cm, H = 1.27 cm, e = 0.952 cm, θ = 17.6°, N = 100 rpm (ω = 0.26 s⁻¹), L = 50.8 cm, M = 30%, and $P_1 - P_2$ = −30.6 atm (−3.10 MPa). Assume Equation 4.22 applies substituting η for μ.

$$Q = G_1 NF_{dt} + \frac{G_2}{\eta} F_{pt}\left(\frac{\Delta P}{L'}\right)$$

Calculate G_1 and G_2.

$$G_1 = \frac{\pi}{2} D^2 H \left(1 - \frac{ep}{\pi D \sin\theta}\right) \sin\theta \cos\theta$$

$$= \frac{\pi^2}{2} (6.35)^2 (1.27) \left[1 - \frac{(0.952)(1)}{\pi(6.35)(0.302)}\right] (0.302)(0.953)$$

$$= 61.2 \text{ cm}^3$$

$$G_2 = \frac{\pi}{12} DH^3 \left(1 - \frac{ep}{\pi D \sin\theta}\right) \sin^2\theta$$

$$= \frac{\pi}{12} (6.35)(1.27)^3 \left[1 - \frac{(0.952)(1)}{(6.35)(0.302)}\right] (0.302)^2$$

$$= 0.261 \text{ cm}^4$$

Estimate F_{dt} and F_{pt}

$$W = \pi D \tan(\theta/p)\cos\theta - e$$

$$= \pi(6.35)(0.317)(0.953) - 0.952$$

$$= 5.07 \text{ cm}$$

$$H/W = 1.27/5.07 = 0.25$$

From Figure 2

$$F_d = 0.87 \qquad F_p = 0.86$$

$$L/D = 50.8/6.35 = 8$$

From Figure 6

$$F_{de} = 1 + \tan^2\Theta - F_{pe}\tan^2\Theta = 0.99$$

$$H/D = 1.27/6.35 = 0.2$$

From Figure 7

$$F_{pc} = 1.095$$

From Figure 8

$$F_{dc} = 1.01$$

From Figure 9

$$F_{pc} = 1.18$$

$$F_{dt} = F_d F_{de} F_{dc}$$

$$= (0.87)(0.99)(1.01) = 0.87$$

$$F_{pt} = F_p F_{pe} F_{pc}$$

$$= (0.86)(1.095)(1.18) = 1.11$$

Estimate η

$$\eta = 78.5\,(\dot{\gamma})^{-0.49}\,\exp(2500/T)\,\exp(-0.079M)$$

$$\dot{\gamma}_H \sim \frac{\pi DN}{H} = \frac{\pi(6.35)(100)}{(1.27)(60)} = 26.2 \text{ s}^{-1}$$

$$\eta = 78.5\,(26.2)^{-0.49}\,\exp[2500/(80+273)]\,\exp[-0.079(30)]$$

$$= 78.5\,(0.202)(1191)(0.0935)$$

$$= 1765 \text{ Ns/m}^2$$

$$Q = \frac{3600 \text{ sec}}{\text{hr}} \left[61.2(100/60)(0.87) + \frac{0.261}{1765}(1.11)\left(\frac{-3.1 \times 10^6}{50.8}\right) \right]$$

$$= 2.83 \times 10^5 \text{ cm}^3/\text{hr} = 0.283 \text{ m}^3/\text{hr}$$

Assuming $\varrho = 1.28 \text{ g/cm}^3$

$$\dot{m} = Q\rho = (0.283)(1280) = 362 \text{ kg/hr}$$

2. Estimate the flow rate in the above problem if H = 0.635 cm.

$$G_1 = 61.2/2 = 30.6 \text{ cm}^3$$

$$G_2 = (0.261)(0.635/1.27)^3 = 0.0326 \text{ cm}^4$$

From the figures

$$F_d = 0.95$$

$$F_p = 0.94$$

$$F_{pe} = 1.095$$

$$F_{de} = 0.99$$

$$H/D = 0.1$$

$$F_{dc} = 1.01$$

$$F_{pc} = 1.08$$

$$F_{dt} = (0.95)(0.99)(1.01) = 0.95$$

$$F_{pt} = (0.94)(1.095)(1.08) = 1.11$$

$$\dot{\gamma}_H = (26.2)(2) = 52.4 \text{ s}^{-1}$$

$$\eta = 7.85(0.144)(1191)(0.0935) = 1256 \text{ Ns/m}^2$$

$$Q = 3600 \left[30.6(100/60)(0.95) + \frac{0.0326}{1256}(1.11)\left(\frac{-3.1 \times 10^6}{50.8}\right) \right]$$

$$= 1.68 \times 10^5 \text{ cm}^3/\text{hr} = 0.168 \text{ m}^3/\text{hr}$$

$$\dot{m} = Q\rho = (0.168)(1280) = 215 \text{ kg/hr}$$

The effects of a non-Newtonian dough on extruder discharge rate can be approximated using Equation 22 developed for Newtonian fluids if η is substituted for μ. η is defined as

$$\eta = \tau/\dot{\gamma} \tag{3.7}$$

and then the relationship between τ and $\dot{\gamma}$ is the power law (Equation 3.5), it follows that $\dot{\gamma}$ is proportional to $\tau^{1/n}$. Since τ is directly proportional to ΔP (Equation 3.16) then

$$\eta = K'\Delta P^{\frac{n-1}{n}} \tag{4.42}$$

Replacing μ in Equation 4.27 with η gives

$$Q = K''\Delta P^{1/n} \tag{4.43}$$

Assuming drag and pressure flows are relatively independent of η then Equation 4.22 can be written

$$Q = G_1 N + \frac{G_2(P_1 - P_2)}{\eta L} \tag{4.44}$$

Combining Equations 42, 43, and 44 and recognizing the ΔPs in Equations 43 and 44 have opposite signs gives

$$Q = \frac{G_1 N}{1 + \dfrac{G_2}{K'K''L}} \tag{4.45}$$

and

$$\Delta P = \left[\frac{G_1 N}{K'' + \dfrac{G_2}{K'L}} \right]^n \tag{4.46}$$

This analysis shows that for a non-Newtonian dough, Q would increase linearly with N but ΔP would increase as N^n. Comparisons of these conclusions with those for a Newtonian dough can be made by examination of Equations 4.28 and 4.29. With a Newtonian dough, both Q and ΔP would be expected to increase linearly with N under isothermal conditions.

3. Flow With Power Law Doughs

As discussed previously, the solution of the equations of motion for a non-Newtonian fluid are very difficult because the flows in the x and z directions are coupled and not independent. Griffith[14] undertook the solution of these equations for a fluid which is described by the power law, considering both the down and cross channel flows. In the development, the temperature dependence on viscosity was also included, but the generalized solution neglected the convective heat transport in the z direction. Consequently, Griffith's results are most pertinent for the isothermal case.

The results of Griffith's numerical solutions have been generalized in Figure 14. The term G_z is the dimensionless pressure flow term considering a power law fluid and is

$$G_z = \frac{P_2 - P_1}{L} \cdot \frac{H^{n+1}\sin\theta}{m(\pi DN)^n} \tag{4.47}$$

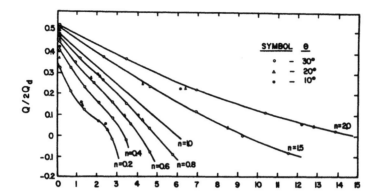

FIGURE 14. Reduced volumetric output vs. $G_z/\cos\theta$ for isothermal flow of power law fluid. (Reprinted with permission from Griffith, R. M., *Ind. Eng. Chem. Fund.*, 1(3), 180, 1962. Copyright © 1962, American Chemical Society.)

where $(P_1-P_2)/L$ = pressure rise along screw, N/m³, H = flight height, m, m = consistency index in Equation 3.7, N s^n/m², n = exponent in power law, D = diameter, m, and N = speed, rps. Q_d has been defined previously when $F_{dt} = 1.0$ as

$$Q_d = G_1 N \qquad (4.48)$$

where G_1 = Equation 4.23 and N = screw speed, rps. Examination of the curves shows that they have greater inflections as n decreases, and the smaller the helix angle at G_z = O, the closer the output tends to pure drag flow. Most importantly, however, the results show that the actual output for a pseudoplastic dough would be less than expected for a Newtonian dough. Application of Figure 14 to data given in Example 1 is shown below.

4. Example 2

Rework Example 1 using Figure 14.

1. For the screw having H = 1.27 cm

$$G_z = \frac{P_2 - P_1}{L} \cdot \frac{H^{n+1}\sin\theta}{m(\pi DN)^n}$$

$$m = 78.5(1191)(0.0935) = 8742 \ Ns^n/m^2$$

$$G_z = \left(\frac{3.1 \times 10^6}{0.508}\right)\left(\frac{(0.0127)^{1.51}(0.302)}{8741\,[(\pi)(0.0635)(1.667)]^{0.51}}\right) = 0.506$$

$$G_z/\cos\theta = 0.506/0.953 = 0.531$$

Reading from Figure 14

$$Q/2Q_d = 0.37$$

From Example 1

$$Q_d = (61.2)(100/60)(3600) = 0.367 \ m^3/hr$$

$$Q = 2(0.367)(0.37) = 0.271 \ m^3/hr$$

The value of Q is approximately 4% below the value calculated in example 1 as would be expected.

2. For a screw having H = 0.635 cm

$$G_z/\cos\theta = (0.635/1.27)^{1.51}\ 0.531 = 0.186$$

From Figure 14,

$$Q/2Q_d = 0.42$$

$$Q = 2(30.6)(100/60)(3600)(0.42) = 0.154\ m^3/hr$$

Difficulties reading Figure 14 can lead to some errors but the flow rate from the extruder is clearly less than that which results from assuming a constant viscosity at the apparent shear rate of the channel.

C. Nonisothermal Newtonian Flow

In the extrusion of food doughs it is common to control the barrel and screw temperatures. The control is achieved by adjusting the steam pressure or electrical power to jackets or heating bands surrounding the barrel and the steam pressure in a hollow-cored screw. In principle it is possible to control the temperature of the dough at the barrel surface and at the root of the screw so that there is a different viscosity in the dough at these two locations. Assuming a linear viscosity profile across the channel depth, Squires and Galt[30] considered one-dimensional flow in the channel, solved the equations of motion (Equation 4.3) in the channel direction, and found the volumetric flow rate to be

$$Q = G_1\,NF_{\mu d}F_{dt} + \frac{G_2}{\mu_m}\,F_{\mu p}F_{pt}\left(\frac{P_1 - P_2}{L}\right) \qquad (4.49)$$

where $\mu_m = (\mu_b + \mu_s)/2$, μ_b = viscosity at barrel surface, μ_s = viscosity at screw root, $F_{\mu d}$ = drag flow viscosity factor, Figure 15, $F_{\mu p}$ = pressure flow viscosity factor, Figure 15, and the other terms have been already defined under Equation 4.22.

Examination of Figure 15 leads to some interesting conclusions. When $\mu_b = \mu_s$, $F_{\mu d}$ and $F_{\mu p}$ are both equal to 1.0 and Equation 4.49 reduces to Equation 4.22. In the case of barrel heating only, $\mu_b < \mu_s$, and the drag flow will be decreased and the pressure flow will increase. In the region where $\mu_d/\mu_s \leq 1.0$, both $F_{\mu d}$ and $F_{\mu p}$ increase. When the screw is heated very hot, the drag flow approaches plug flow with the expected doubling of the rate. In the narrow region $1 \leq \mu_b/\mu_s \leq 10$, $F_{\mu d}$ increases more rapidly than $F_{\mu p}$. Since drag flow usually predominates in food extrusion, heating of the screw will normally result in an increase in extruder output and careful attention to the relative temperatures of the screw and barrel are not critical as long as the screw temperature is at least equal to the barrel temperature.

D. Nonisothermal Non-Newtonian Flow

In a cooking extruder, flow would have to be characterized as nonisothermal and non-Newtonian. The temperature of the food dough is influenced by jackets surrounding the extruder barrel and the extrusion screw which can be either heated or cooled. In addition, the internal generation of heat within the dough from the viscous dissipation of the mechanical shaft energy can cause significant increases in the temperature of the food product. In some cooking extruders, steam and/or water are injected di-

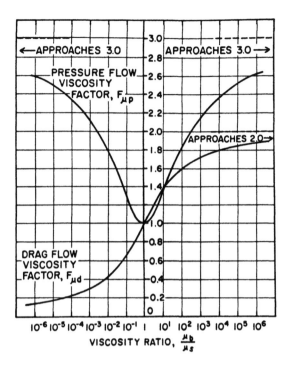

VISCOSITY RATIO, $\frac{\mu_b}{\mu_s}$

FIGURE 15. $F_{\mu d}$ and $F_{\mu p}$ as a function of μ_b/μ_s. (From Paton, J. B., Squires, P. H., Darnell, W. H., Cash, F. M., and Carley, J. F., *Processing of Thermoplastic Materials*, Bernhardt, E. C., Ed., Robert E. Kreiger, Huntington, N.Y., 1974, 186. With permission. © 1974 Society of Plastics Engineers, Inc.)

rectly into the barrel and the resulting changes in temperature and moisture greatly affect velocity and viscosity distributions.

The development of an analytical model in closed form which adequately handles all these contingencies has not occurred because of the mathematical complexities in formulating and solving the problem. Several papers have been published[9,14] which made the assumption that there is no convective or conductive heat transfer in z direction within the channel which has to be seriously questioned, considering the nature of the flow within the channel and the possibility that the barrel and screw surfaces are at different temperatures.

In the thermoplastics industry, computer simulations of the metering section of an extrusion screw containing temperature sensitive non-Newtonian polymer melts have been performed.[11] These simulations require that the channel be shallow in the metering section and considers temperature increases due to viscous dissipation of mechanical energy and thermal conduction across the depth of the channel. For computational reasons, the analysis considers the velocity in the z down channel direction only and neglects the thermal convection due to transverse flow in the channel.

Applications of the computer simulations of these complex flow conditions would require a thorough knowledge of the properties of the food product, including its rheology, thermal conductivity, specific heat, density, etc., as well as the heat transfer coefficients at the barrel and surface of the screw. For most food systems, these are not known which leads to significant problems in applying the theoretical developments made in the area of thermoplastics where the properties of the materials are reasonably well characterized. The cooking or other chemical reactions which also oc-

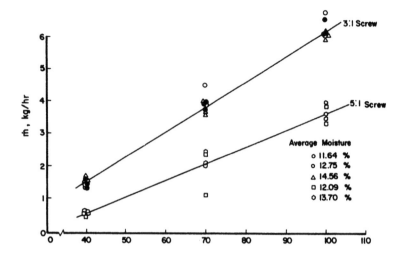

FIGURE 16. Flow vs screw speed for corn grits extrusion. (From Harmann, D. V. and Harper, J. M., *Trans. ASAE*, 16(6), 1175, 1973. With permission.)

cur within food systems during extrusion further alters physical properties and complicates the problem.

Additional consideration of the non-Newtonian, nonisothermal flow equations is beyond the scope of this book. The interested student is referred to the in depth treatment of the subject in the books by Fenner[12] and Tadmor and Klein.[32]

E. Application of Flow Models

Despite the many complexities when biopolymeric food systems are extruded, the application of the simplified extrusion theory has been successfully applied and provides considerable insight into the measured effects of changing screw geometry and extruder operating conditions. Most of the pertinent literature in this area is cited below as examples.

Griffith[14] used corn syrup as a model Newtonian fluid in a 5.08-cm diameter, Hartig No. 9 plastic extruder with a single flighted screw having $\theta = 30°$ and H>0.32 cm. The extruder was jacketed and had a hollow screw which allowed for the circulation of water to maintain isothermal conditions. Actual flow measurements were about 8% below those estimated by an equation similar to Equation 4.22.

Harmann and Harper[15] reported flow rates measured on a 1.90-cm, Brabender® extruder as a function of screw speed, compression ratio, and moisture content of corn grits. Some data from these studies are shown in Figure 16. It is clear from these studies that flow increases linearly with screw speed, N, as predicted in Equation 4.22 and is little influenced by moisture. The value of G_1 (Equation 4.23) was, however, nearly double the value actually measured in the experiments with the difference being attributed to leakage flow which occurred through longitudinal grooves in the barrel and the pressure flow within the channel. Pressure flow down the grooves in this study would have had to be about one-eighth of that predicted by Equation 4.37 considering the entire groove area indicating the grooves were partially filled with stagnant solids, a condition which was confirmed through visual observations. The ratio of the geometric drag flow constants, G_1, calculated for the two screws tested was nearly equal to the ratio of flows actually measured at a constant N. These data would indicate that the ratio of G_1N for different screws would be an effective means of predicting flow from extruders where pressure flow does not change significantly.

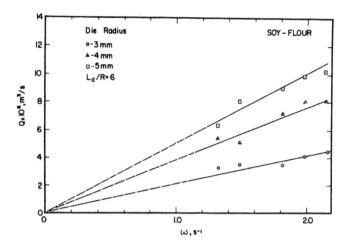

FIGURE 17. Experimental flow rates as a function of screw speed for different dies. N, rpm = 9.55 ω. (From Bruin, S., van Zuilichem, D. J., and Stolp, W., *J. Food Proc. Eng.*, 2(1), 1, 1978. With permission. © Food and Nutrition Press, Inc.)

A similar study reported by Bruin et al.[4] compared flows of extruded corn grits through a Bottenfield® extruder with flows predicted by Equation 4.22. These researchers found that the theory predicted a flow rate somewhat higher than actually measured and that deviations became higher with screws having lower compression ratios. The same researchers have also experimentally measured flow rates of modified di-amylopectin phosphates and defatted soy grits with varying die configurations. The data for the soy grits extrusion are presented in Figure 17. The samples had moisture contents ranging from 19 to 27%$_w$ and pressures at the die were in the region of 0.25 to 0.70 MPa. Despite the complex rheological behavior of the food dough, these data again indicate that flow rate, Q, increased linearly with N as predicted by Equation 4.22. The example below also shows how these data consistently reflect the influence of the die diameter on the extrusion output.

1. Example 3

A reduced extrusion rate is shown on Figure 17 at constant ω or N when smaller dies are used. Show that the reductions measured are consistent with simplified extrusion theory.

Assumption: viscosity of defatted soy dough in the channel of the screw is constant at equal screw speeds and extrusion temperatures, the metering section of the screw controls flow and end effects on the die can be neglected.

Extrusion flow can be represented by

$$Q = G_1 N F_{dt} + \frac{G_2}{\mu} F_{pt} (\Delta P/L) \qquad (4.22)$$

where for the purposes of calculation F_{dt} and F_{pt} are assumed equal to 1. The difference in flow with two die sizes at constant N is

$$Q' - Q'' = -\frac{G^2}{\mu L} (\Delta P'' - \Delta P')$$

where the primes denote different die sizes.

The flow through a die hole is related to ΔP by

$$Q = K \frac{\Delta P}{\mu_d} \qquad (4.27)$$

where $K = \pi R^4/8L_d$ for round dies, R is the hole radius, and L_d is its length. For the data under consideration, $L_d = 6R$ for all dies so

$$K = f(R^3)$$

Soy doughs are pseudoplastic and their viscosity can be represented by

$$\eta = f(\dot{\gamma})^{n-1} \qquad (3.8)$$

where an average value of n is 0.3 (see Table 3 in Chapter 3) and

$$\dot{\gamma}_w = \frac{3n+1}{4n}\left(\frac{4Q}{\pi R^3}\right) = f\left(\frac{Q}{R^3}\right) \qquad (3.42)$$

resulting in

$$\eta = f(Q/R^3)^{0.3-1}$$

Substituting the values for K in terms of R and $\eta = \mu_d$ from above into Equation 4.27 and rearranging gives

$$\Delta P = f(Q/R^3)^{0.3}$$

Using this value, ΔP can be eliminated from the equation showing differences in flow giving

$$Q' - Q'' = -\frac{G_2}{\mu L} f\left[(Q''/R''^3)^{0.3} - (Q'/R'^3)^{0.3}\right]$$

or

$$\frac{(Q' - Q'')}{[(Q''/R''^3)^{0.3} - (Q'/R'^3)^{0.3}]} = -G$$

where G is some constant.

The consistency of the data can be tested with this equation for the three die sizes as shown in the table below.

R, mm	$Q \times 10^6$ m^3/s^{-1} $\omega = 2.0$ s^{-1}	$(Q/R^3)^{0.3}$	Dies for which flow data are compared	G
3	4.0	0.56	5 mm & 3 mm	57
4	7.5	0.52	5 mm & 4 mm	37
5	9.7	0.46	4 mm & 3 mm	88

For perfect consistency all Gs would be constant for all die combinations, which

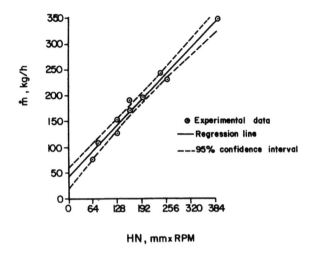

FIGURE 18.   Extrusion rate as affected by flight depth (mm)
in metering section times screw speed (rpm). (From Tsao, T.
F., Harper, J. M., and Repholz, K. M., *AIChE Symp. Ser.*,
74(172), 142, 1978. With permission.)

obviously is not the case. Nonetheless, the relatively constant Gs calculated are sup-
portive of the validity of applying simplified extrusion theory to complex situations to
estimate trends even though a large number of assumptions are required.

In another study reported by Harmann and Harper,[16] pregelatinized corn flour hy-
drated to 32% moisture was isothermally extruded at 88°C. A Brabender® extruder
was used having both 1.9- and 3.18-cm smooth barrels with 20:1 L/D ratios. Five
screws having varying compression ratios were also used. Knowing the dough viscosity
and pressure profiles, experimental measurements for flow ranged from 9 to 13%
above predicted flow for shallow screws. For one deep flighted screw, theory overpredicted flow by 21%. Deviations from theory were attributed to the non-Newtonian
behavior of the dough which was only partially accounted for by substituting η for μ
in the simplified model.

A series of studies on 14 specially designed extruder screws with varying flight
depths, pitches, compression ratios, and lengths of metering section was conducted by
Tsao et al.[33] The feed material was a high moisture food with a relatively low viscosity,
which resulted in low pressure drops across the screw. Under these conditions, the drag
flow term in Equation 4.22 was the dominant factor in the flow equation. Figure 18
shows extruder output varying linearly with HN as predicted. Generally, the flow rate
of shallow flighted screws was underpredicted while the flow for deep flighted screws
was overpredicted. These deviations were attributed to the non-Newtonian viscosity of
the food material altering the velocity profile within the channel. Their data also
showed that Equation 4.22 underpredicted flow for screws having a low pitch angle.

Equation 4.40 for flow in extruders equipped with screws having tapered channels
was used in the analysis of data in the study by Harmann and Harper[16] and Tsao et
al.[33] The equation tended to underpredict flow by about 10% for the cases studied.

Using the data of Mustakas et al.[24] on the extrusion of full-fat soy flour, Fricke et
al.[13] used the model for the metering section of the extruder, including heating effects
from the jackets of the barrel and viscous dissipation. Their results showed flow was
not affected by the pressure rise in the extruder, meaning drag flow dominated. At
low screw speed, flow was proportional to screw speed as anticipated. At higher screw
speeds, a great deal of data scatter occurred and was attributed to surging in the extruders screw.

From the above examples it appears that extrusion theory can be a valuable tool in predicting flow in food extrusion but additional work is needed to account for all phenomena occurring in cooking extrusion.

III. POWER

A. Power Inputs

The rotation of the extruder screw provides input of mechanical energy. Through viscous dissipation, most of this energy is converted to heat in the dough but some goes to increase the pressure in the food dough and its kinetic energy. Estimation of power inputs allows the calculation of drive power requirements, an essential parameter in extruder specification. The following is the development of the equations for the power inputs into the metering section of the screw where the most significant power input occurs.

The total power input to the shaft of an extruder can be expressed as

$$dE = dE_H + dE_p + dE_k + dE_\delta \qquad (4.50)$$

where dE = energy input per differential down channel distance, dE_H = viscous energy dissipation in channel, dE_p = energy to raise pressure of fluid, dE_k = energy in increase kinetic energy, and dE_δ = viscous energy dissipation in flight clearance. Each term will be examined separately below. Normally, since velocities in an extruder are low, dE_k is assumed negligible.

1. Power Dissipated in Channel

The power input into food material in the channel element WHd_z is written

$$dE_H + dE_p = p \left[\int_O^W \tau_{yz}|_{y=H} V_z dx + \int_O^W \tau_{yx}|_{y=H} V_x dx \right] dz$$
$$(4.51)$$

where the shear stress, τ, and velocity have the same direction. Since no flow occurs in the y direction, the power requirement for this component is zero. Shear stresses are related to viscosity as follows

$$\tau_{yz}|_{y=H} = \mu(\partial v_x/\partial y)_{y=H} \qquad (4.52)$$

$$\tau_{yz}|_{y=H} = \mu(\partial v_z/\partial y)_{y=H} \qquad (4.53)$$

It is necessary to evaluate the shear rate in the x and z directions at the barrel surface which can be done by differentiating the velocity profiles in Equations 4.14 and 4.20. Evaluating these derivatives at y = H gives

$$\partial v_x/\partial y|_{y=H} = -4V_x/H \qquad (4.54)$$

and

$$\partial v_z/\partial y|_{y=H} = \frac{V_z}{H} + \frac{H}{2\mu}\left(\frac{\partial P}{\partial z}\right) \qquad (4.55)$$

all of which can be substituted in Equation 4.51 and following integration gives

$$\frac{dE_H}{dz} + \frac{dE_p}{dz} = p\left(\mu V_z^2 \frac{W}{H} + \frac{V_z}{2} WH \frac{\partial P}{\partial z} + 4\mu V_x^2 \frac{W}{H}\right)$$

(4.56)

Knowing that

$$V_x = \pi ND \sin \theta$$

and

$$V_z = \pi ND \cos \theta$$

gives

$$dE_H + dE_p = p\mu(\pi ND)^2 \frac{W}{H} (\cos^2\theta + 4\sin^2\theta)dz + p\pi ND \frac{WH}{2} \frac{\partial P}{\partial z} \cos\theta dz \qquad (4.57)$$

The viscous dissipation of energy in the channel is proportional to μ and N^2. The energy requirement for the pressure increase is directly proportional to N, which is related to Q and the pressure increase along the channel. In most cases, the energy to increase the pressure on the dough is very small compared to the viscous dissipation term due to the relatively low ΔPs and high μs involved.

2. Power Dissipated in Flight Clearance

The power input into an element of dough δedz in the flight clearance due to drag flow will be

$$dE_\delta = p\left[\int_O^e \tau Vdx\right] dz$$

(4.58)

where the shear rate $\partial v/\partial y$ is approximately V/δ and the shear stress is

$$\tau = \mu_\delta \frac{V}{\delta}$$

(4.59)

The viscosity in the clearance is μ_d to denote the fact that it may be significantly different from the viscosity in the channel because of temperature and shear effects. Substituting the value for τ into Equation 4.58 gives

$$dE_\delta = p\mu_\delta \left[\int_O^e \frac{V^2}{\delta} dx\right] dz$$

(4.60)

Using $V = \pi DN$ and integrating gives

$$dE_\delta = p\mu_\delta \frac{(\pi DN)^2 e}{\delta} dz$$

(4.61)

Equation 4.61 shows the high significance of the clearance between the screw and the barrel on power input of cooking extruders with small δs leading to larger power inputs. The off-setting factor is that when δ is small, μ_δ is small.

3. Total Power Input to Screw

Adding the various contributions to total power input to the screw gives

$$dE = dE_H + dE_p + dE_\delta$$

$$dE = p\left[\mu(\pi ND)^2 \frac{W}{H}(\cos^2\theta + 4\sin^2\theta) + \pi NDW \frac{H}{2}\frac{\partial P}{\partial z}\cos\theta\right.$$

(4.62

$$\left. + \mu_\delta(\pi ND)^2 \frac{e}{\delta}\right] dz$$

(4.63)

Recognizing that

$$dL = dz\sin\theta$$

(4.64)

and integrating over the entire length of screw, L, gives total power input as

$$E = p\frac{(\pi ND)^2 L}{\sin\theta}\left[\mu\frac{W}{H}(\cos^2\theta + 4\sin^2\theta) + \mu_\delta\frac{e}{\delta}\right]$$

$$+ p\frac{\pi NDWH}{2}\Delta P\cos\theta$$

(4.65)

Equation 4.65 is sometimes rewritten as

$$E = G_s N^2\left[\mu\frac{W}{H}(\cos^2\theta + 4\sin^2\theta) + \mu_\delta\frac{e}{\delta}\right] + G_1 N\Delta P$$

(4.66)

where

$$G_s = \frac{p(\pi D)^2 L}{\sin\theta}$$

(4.67)

and

$$G_1 = \frac{\pi}{2}D^2 H\left(1 - \frac{ep}{\pi D\sin\theta}\right)\sin\theta\cos\theta$$

(4.68)

Input torque requirements can be calculated by dividing Equation 4.66 by N.

Specific power, $E/\dot{m}$, is the power per unit flow rate and is a common way of expressing power requirements. Since in food extrusion flow rate is nearly equal to the drag flow or $G_1 N$, specific power would be expected to increase linearly with N for Newtonian doughs at constant T. For pseudoplastic doughs, the viscosity is proportional to $\dot{\gamma}$ which is in turn proportional to N^{n-1}, so specific power would be expected to increase as N^n.

Strictly speaking, Equations 4.65 or 4.66 describe the power inputs for an extruder operating with Newtonian fluids only. These equations can be used to approximate the power inputs for non-Newtonian fluids if η is substituted for μ. In this case, η needs to be evaluated at $\dot\gamma$ (Equation 4.41) and temperature in the channel. Likewise, the viscosity in the clearance needs to be evaluated at $\dot\gamma_d = V/\delta$ and the temperature in the clearance.

Following Bruin et al.[4] to eliminate ΔP in the above equation, the results of Equation 4.29, which gives ΔP to be a function of extruder and die geometry for a particular dough type, may be substituted above. Dividing both sides of Equation 4.65 by $\mu p N^2 D^3$ gives a power number similar to that used in the analysis and design of mixing systems in the viscous flow region.

$$\frac{E}{\mu p N^2 D^3} = \frac{\pi^2}{\sin\theta}\frac{L}{D}\left[\frac{W}{H}(\cos^2\theta + 4\sin^2\theta) + \frac{\mu\delta}{\mu}\frac{e}{\delta}\right]$$

$$+ \frac{\pi WH}{2\mu D^2}\left(\frac{\dfrac{G_1 F_{dt}}{\dfrac{K}{\mu_d} + \dfrac{G_2 F_{pt}}{\mu L}}}{}\right)\cos\theta \qquad (4.68A)$$

It should be noted that the power number is a constant for a given extruder and die design.

4. Example 4

Calculate the power input to the metering section of the two screws used in Example 1 with a $\delta = 0.050$ cm.

1. Calculate power requirements for $H = 1.27$ cm. Equation 4.66 can be used to calculate power inputs if η is substituted for μ and evaluated at the appropriate temperature and $\dot\gamma$s. Assume that the temperatures in the clearance and the channel are 80°C. This is a questionable assumption since most of the energy is dissipated at the barrel of the screw and implies that rapid heat transfer exists at this point to maintain a constant dough temperature. Since 80°C is undoubtedly cooler than the actual temperatures at these locations, the equations will tend to overestimate the power requirements.

$$\dot\gamma_\delta = \frac{\pi DN}{\delta} = \frac{\pi(6.35)}{0.050}\frac{100}{60} = 664\ s^{-1}$$

$$\eta_\delta = 78.5(664)^{-0.49}(1191)(0.0935) = 362\ Ns/m^2$$

From Example 1

$$\eta = 1765\ Ns/m^2$$

and

$$G_1 = 61.2\ cm^3 = 6.12 \times 10^{-5}\ m^3$$

$$G_8 = \frac{p(\pi D)^2 L}{\sin\theta} = \frac{1(6.35\pi)^2}{0.302}(50.8) = 66{,}900\ cm^3 = 0.0669\ m^3$$

$$E = (0.0669)(100/60)^2\left[1765\frac{5.07}{1.27}(0.953)^2 + 4(0.302)^2) + 362\left(\frac{0.952}{0.05}\right)\right]$$

$$+ 6.12 \times 10^{-5}(100/60)(3.1 \times 10^6)$$

$$= 2950 + 316 = 3264\ Nm/s = 3.27\ kW$$

Use an 11-kW (15 hp) motor to take into account the energy used in the feed and transition sections and to accommodate surges. The example uses a relatively nonviscous dough which requires low mechanical energy input.

2. Calculate power requirements for H = 0.635 cm.

From Example 1

$$\eta = 1256 \ Ns/m^2$$

and

$$G_1 = 30.6 \ cm^3 = 3.06 \times 10^{-5} \ m^3$$

$$E = (0.0669)(100/60)^2 \left[1256 \frac{5.07}{0.635}(0.953)^2 + 4(0.302)^2) + 362\left(\frac{0.952}{0.05}\right) \right]$$

$$+ \ 3.06 \times 10^{-5} (100/60)(3.1 \times 10^6)$$

$$= 3653 + 158 = 3811 \ Nm/s = 3.81 \ kW$$

The 11 kW (15 hp) motor specified for the deeper screw is still sufficient for the shallow screw.

B. Application of Power Model

References to the application of the models describing the power inputs to a food extruder are relatively few. Total power inputs to the extruder are commonly measured with a kW meter on the electrical drive motor. Losses of power through the drive train including speed reducers and bearings need to be subtracted from total power if proper correlation is to be made with the power calculated with Equation 4.66 for the metering section of the extrusion screw. To apply Equation 4.66, it is necessary to have a thorough understanding of the rheology of the food materials as a function of temperature. Since most food doughs are non-Newtonian, it is appropriate that the rheology be known as a function of $\dot{\gamma}$.

Harmann and Harper[16] calculated the torque requirements for extruding a rehydrated pregelatinized corn dough and compared the results with actual measurements. Their experiments were conducted isothermally and the rheology of the food dough was correlated with the power law which was evaluated at $\dot{\gamma}$ estimated by Equation 4.41. Their data showed that the torque estimates made with Equation 4.66 divided by N, were 17 to 73% higher than the values actually measured. The inability of the equation to predict the actual value was attributed to the uncertainty in knowing the temperature of the dough at the barrel surface and the active or filled length of the extrusion screw. Since the temperature at the barrel surface was probably higher than the bulk temperature of the dough and the active length less than the entire screw length, the equations would naturally overpredict the power requirements.

Several references have been made to power requirements of corn grits extrusion.[4,15] These studies showed power requirements decreasing with increasing dough temperature and moisture content. Over the range of conditions studied, specific powers of 0.10 to 0.13 kW-hr/kg were found. The results of Bruin et al.[4] were plotted as power number (Equation 4.68) against Reynolds number (Re = pND^2/μ), where μ, the viscosity, was taken as that of corn grits dough at a moisture, temperature, and shear rate at the screw tip. The resulting data for each moisture dough was constant, as would be anticipated. Higher moisture showed lower power numbers, and the reason that data for different moistures did not coincide on a single curve was attributed to

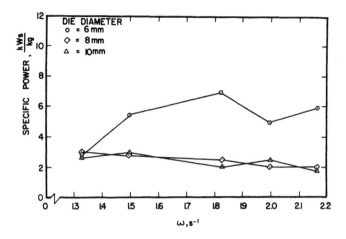

FIGURE 19. Specific power consumption during extrusion of defatted soy flour. N, rpm = 9.55 ω. (From Bruin, S., van Zuilichem, D. J., and Stolp, W., *J. Food Proc. Eng.,* 2(1), 1, 1978. With permission. © Food and Nutrition Press, Inc.)

the lack of more specific knowledge about the viscosity of the corn grits dough. As moisture contents of the corn grits increases, additional starch gelatinization can occur, causing changes in the dough viscosity.

The effect of increasing extrusion speed on specific power was measured by Tsao et al.[33] in experiments with 14 different screws using a high moisture food dough. Their data showed specific power increasing approximately as N^n for a threefold, screw-speed increase, as expected. Actual power requirements were less than those calculated and could be attributed to the actual dough temperatures being higher than measured at the barrel surface.

A number of specific power measurements are given for mixtures of cereal grains with 0 to 32% whole soy for a simple autogenous extruder at moistures of 12 to 20%ₒ by Harper.[18] These data show ranges of 0.08 to 0.16 kW-hr/kg, with wheat and sorghum blends showing higher values than did corn or rice. These data showed that the specific power did not increase with N as anticipated. Bruin et al.[4] also found that specific power did not increase with N in the extrusion of defatted soy flour as shown in Figure 19. Several things may account for these findings, including the need for a more accurate understanding of the effects of moisture and shear rate on viscosity and a correct measurement of the temperature of the dough.

Bruin et al.[4] showed data on specific power requirements with a modified diamylopectin phosphate dough at different moistures, with an extruder equipped with different dies and screws at a single screw speed. Their data showed reduced specific power with increasing moisture, but also increasing at higher moisture. They attributed these results to increased gelatinization at higher moistures.

An increasing specific power with decreasing H is clearly shown in Figure 20 from the work by Tsao et al.[33] For a Newtonian dough, the increase in specific power with decreasing H would have been more apparent. For the case studied, the increased $\dot{\gamma}$ in the screw channel at reduced Hs resulted in lower viscosities and a lessened effect.

C. Energy Balance

The previous section dealt with the power required to drive the extrusion screw, provide the viscous energy dissipated in the channel and flight clearance, and increase the pressure on the product. During the extrusion process, this mechanical energy is

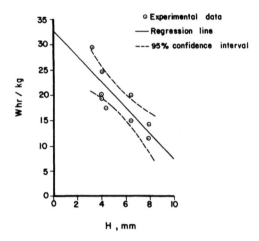

FIGURE 20. Specific power vs. flight depth at N = 60 rpm. (From Tsao, T. F., Harper, J. M., and Repholz, K. M., *AIChE Symp. Ser.*, 74(172), 142, 1978. With permission.)

principally dissipated in the form of heat which causes temperature, chemical, and perhaps phase (latent heat) changes in the food product. The amount of energy dissipated may be improperly matched with the extent of the thermal requirements. In this case, heat can be either added or removed through jackets around the barrel and/or a cored screw having a quill through which either water or steam are circulated. Direct steam injection to the food in the extrusion screw can also be another way to make up a heat deficit.

An energy balance around the extruder provides a clearer understanding of the energy inputs and their distribution as described above. Simply, the energy balance per unit of time is written:

$$\frac{E_t}{\Delta t} = \frac{E}{\Delta t} + q \qquad\qquad (4.69)$$

and

$$\frac{E_t}{\Delta t} = Q\rho \left[\int_{T_1}^{T_2} c_p dT + \int_{P_1}^{P_2} \frac{dP}{\rho} + \Delta H^\circ + \Delta H_{s_1} \right] \qquad (4.70)$$

where E_t = total net energy added to extruder, E = mechanical energy dissipated (from Equation 4.65), Δt = time interval for energy balance, q = heat flux to dough (+) or loss (−), $Q\rho$ = $\dot{m}$, mass rate of flow, c_p = specific heat, T = temperature, P = pressure, ΔH° = heat of reaction/unit mass dough, endothermic (+), ΔH_{s_1} = latent heat of fusion/unit mass of dough, and subscripts 1 = feed port of extruder and 2 = just behind the die. The heats of reaction can account for chemical changes occurring in the food such as starch gelatinization, protein denaturation, browning, etc. Their reactions will be discussed in greater detail in Chapters 11 and 13 dealing with the extrusion of starch and protein-based foods. In the energy balance the heats of reaction are based on a unit mass of food. It is conceptually easier to think of this term as being the net heat of reaction for all reactions which occur in the food system. Sahagun and Harper,[28] using blends of 70% corn/30% soybean, found that approximately 15% of the total energy added to the autogenous extruder was unaccounted for by sensible heat changes, pressure increases, and heat losses. They attributed this difference to heats of reaction associated with the chemical changes occurring in the food materials.

The latent heat of fusion is also included in the energy balance to account for a relatively small amount of energy associated with the melting of solid lipid materials that may be part of a food formulation. Because the boundary for the balance considers the food before it emerges from the die, no latent heat associated with the formation of steam at the extruder discharge needs to be considered.

IV. COMPLETE MODELS

Until now, the extrusion models discussed were exclusively developed for the metering section of the screw. Since the metering section is often rate limiting and the region where a large portion of the power is dissipated, it is appropriate that the emphasis of most of these chapters be focused there. However, brief discussions of the feed section and transition sections of the extrusion screw are in order to give a complete picture of the status of extrustion models for food extrusion.

A. Feed Section

In the feed section of the extruder, relatively free-flowing granular particles of the food ingredients exist. This granular material is caught between the flights of the screw channel and is conveyed in a manner similar to the action of a screw conveyor. Little or no internal shear of the food takes place in the solids conveying section as contrasted to shear flow in the metering section. The granular material acts like a solid plug contacting the screw channel on all sides.

The coefficient of friction between the food and the screw and barrel are important in evaluating the flow of a solid plug. The angle of movement of the outer surface of the solid plug has been shown by Tadmor and Klein,[32] using a force balance, to be

$$\cos \theta_\sigma = K \sin \theta_\sigma + \frac{f_s}{f_b} \sin \theta \ (K + \cot \theta) \left(1 + \frac{2H}{W}\right)$$
$$+ \frac{H}{Z} \frac{1}{f_b} \sin \theta \ (K + \cot \theta) \ln \frac{P_2}{P_1}$$

(4.71)

where θ_o = angle of movement of outer surface of solid plug, θ = helix angle of feed section, f_s = coefficient of friction — food and screw, f_b = coefficient of friction — food and barrel, H = flight height in feed section, W = width of channel perpendicular to flight, Z = length of channel, P = pressure, and

$$K = \frac{\sin \theta + f_s \cos \theta}{\cos \theta - f_s \sin \theta}$$

(4.72)

or the ratio of forces acting in the solids bed. Unless there is specific information to the contrary, it is often assumed that $f_s = f_b$.

Equation 4.71 can be rewritten as

$$\cos \theta_\sigma = K \sin \theta_\sigma + M$$

(4.73)

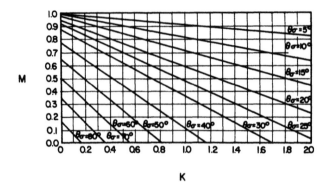

FIGURE 21.   Relationship between M, K, and θ_o. (From Darnell, W. H. and Mol, E. A. J., *SPE J.*, 12, 20, 1956. With permission.)

where M is the last two terms in Equation 4.71. The information given in Figure 21 allows the calculation of θ_o directly once K and M are calculated.

Knowing θ_o, the solids flow rate can be calculated by

$$Q_\sigma = \pi^2 NHD(D-H)\left[\frac{\tan\theta_\sigma \tan\theta}{\tan\theta_\sigma + \tan\theta}\right]\left(\frac{W}{W+e}\right) \qquad (4.74)$$

where Q_o = volumetric flow of solids, D = screw diameter, N = screw speed, and e = width of flight measured perpendicular to flight face. Examination of Equation 4.74 indicates that Q_o is increased by a feed section having deep flights (large H), when there is a low coefficient of friction between food and screw, a high coefficient between the food and barrel, and by increasing N. For optimal solids conveying, it is common to use a polished screw and a barrel which is grooved to prevent slip at the surface.

It is not uncommon for food extruder screws to have deep feed sections resulting in compression ratios of three or more. The deeper flights more readily accept the food ingredients and begin to convey them assuming that the metering section is not starved. The feed material typically has a lower bulk density than does the compacted dough, making greater flight depths necessary. The optimal helix angle for food materials is between 10 and 20°. For wet sticky materials, lower helix angles often result in better conveying in the feed section of the screw.

The coefficients of friction between food materials and extrusion surfaces are not widely reported. The ASTM method D1894-63[1] for plastic materials was used successfully by Jasberg et al.[20] and their data for soy grits is shown in Figure 22. These authors also found the theory for solids conveying gave good agreement with experimental results.

B. Transition Section

The transition section of a food extrusion screw is where the food ingredients change from a raw or uncooked granular material to the plastic-like dough which exists in the metering section of the screw. In the extrusion of thermoplastics, the transformation of granular particles of resin to a hot fluid material is called melting and occurs principally in the transition of the screw. Although superficially the changes that occur within the food ingredients appear similar to the melting of the thermoplastic resin, the only thing these two processes share is the similarity of the physical appearance of the materials before and after the transition section.

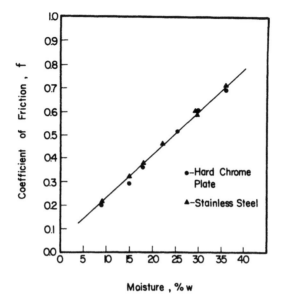

FIGURE 22. Coefficient of friction between soy flakes and metal surfaces. (From Jasberg, B. K., Mustakas, G. C., and Bagley, E. B., Pres. Soc. Rheology, 48th Annu. Meeting, Madison, Wis., 1977. With permission. © 1977 Society of Plastics Engineers.)

In the case of thermoplastics, the transition from solid to a fluid is described as melting which is associated with a latent heat of fusion, and the resulting material is properly called a melt. In the food system, this change of physical character is associated with a set of chemical reactions called cooking for lack of a more descriptive term. Cooking encompasses the following thermal induced changes:

1. Hydration and denaturization of protein
2. Hydration, gelatinization, and dextrinization of starch
3. Browning where free ε-amino groups of the protein react with reducing sugars
4. Denaturization of antigrowth factors, vitamins, and enzyme systems which occur in raw foods
5. Destruction of microorganisms which exist in the raw food

Simply, cooking involves the conversion and/or reaction of the major food constituents — carbohydrate, fat, protein, and water. In this context, the application of the words melt or melting to food extrusion is a misnomer.

During the cooking of the food that begins in the transition section and proceeds into the metering section of the screw, it is not uncommon to see the viscosity and other physical properties of the dough change rather drastically. Cooking can cause an increase in viscosity while melting is associated with decreased viscosity. The modeling of these complex phenomena in food extrusion has never been undertaken. Basic data that describe the cooking of food under low- to intermediate moisture conditions is not available. Moreover, the effect of cooking on the physical properties of food during the cooking process is unknown.

In the case of thermoplastics extrusion, the melting which occurs in the transition zone of the screw has been extensively studied. As the bed is worked and compressed in the transition zone, melting occurs at the barrel surface and this material flows to

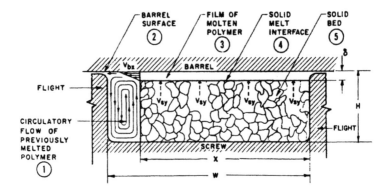

FIGURE 23.   Transition channel cross section showing melting of thermoplastic polymer. (From Tadmor, Z. and Klein, I., *Engineering Principles of
Plasticating Extrusion*, Van Nostrand Reinhold, New York, 1970, 108. With
permission. © 1970 Society of Plastics Engineers, Inc.)

form a pool of melted polymer at the rear of the flight. Such a process is shown in
Figure 23 where the size of the pool of melted polymer increases as the material moves
through the screw. As discussed previously, the similarity between thermoplastic materials and foods is tenuous but when the screw of a food extruder is cooled and removed, the pattern of materials within the flight is similar to that shown in Figure 23.
Uncooked and granular food material exists in the forward part of the flight in more
or less a solid bed, and the cooked doughy material accumulates at the rear of the
flight.

Modeling of the melting process which occurs in the case of thermoplastics consists
of a heat balance around an element within a flight, where heat is generated within
the element by viscous dissipation and can be conducted into the element from the
barrel or screw. The effect of this heat on melting and viscosity of the material within
the element can be related to the relative portions of solid or melted polymer within a
flight — the power consumed, temperature, and pressure developed as a function of
screw speed, N, barrel temperature, flight clearance, screw geometry, etc. For more
information on the subject the reader is referred to Tadmor and Klein.[32]

Such modeling techniques have been powerful tools assisting in the optimal design
of screws for the extrusion of thermoplastics. Once a better understanding of the cooking of food is achieved, it appears that more advantage may be taken of the extensive
work in thermoplastics to optimize the design of the food extrusion screw. Foods are
natural materials that have considerable variability, so a precise understanding of the
manner in which physical properties (such as specific heat, c_p, density, ϱ, thermal conductivity, k, viscosity, η, as well as chemical properties change as a function of temperature, time, moisture, and shear environment, will be difficult. Despite these difficulties, which greatly surpass the complexities of thermal-plastic systems, further focus
on the transitions which occur in food systems during cooking extrusion appear very
fruitful. Once these are known, it becomes theoretically possible to completely model
mathematically the food extruder, coupling the models for the feed, transition, and
metering sections so that more valid estimates of flow, power, and heating requirements and product characteristics may be made.

V. DIES

A. Die Characteristics

The extruder die is the shaped hole through which the food dough flows at the extruder exit. The cross-sectional area of the die gives the shape to the extruded food product. The construction and shape of the die, necessary to give uniform flow and correct dimensional size, requires a great deal of experience and in certain cases, trial and error when the shape is elaborate. The basic relationships between the characteristics of the die design and flow and pressure drop are given below.

For Newtonian fluids, the flow through various shapes, neglecting entrance and exit losses, has been previously given as

$$Q = K \frac{\Delta P}{\mu_d} \qquad (4.27)$$

which shows Q directly proportional to ΔP and inversely proportional to μ_d. For some common cross sections, K is given as (Paton et al.[25])

circle: $K = \pi R^4 / 8L_d$ (4.75)

slit: $K = 2wC^3 / 3L_d$ (4.76)

annulus: $K = \dfrac{\pi(R_o + R_i)(R_o - R_i)^3}{12L_d}$ (4.77)

where R = radius, L_d = length of die, w = width of slit, 2C = height of slit, and o,i = subscripts for outside and inside. Notice that the K for an annulus would be calculated for a slit having a width equal to $\pi(R_o + R_i)$ and a height of R_o-R_i which is approximately correct for annuli that are very narrow.

Most food doughs are highly non-Newtonian as discussed in Chapter 3. In these cases, viscosity is a function of shear rate and so a nonlinear relationship will exist between Q and ΔP. To do die design for non-Newtonian food doughs or evaluate die characteristics, it is necessary to have some relationship between viscosity and shear rate such as Equation 3.8, the power law model.

To estimate the ΔP through a die at a given Q for a non-Newtonian dough, the shear rate, $\dot\gamma$, and the shear stress, τ, are first calculated. For various cross sections, $\dot\gamma_w$ and τ_w are

circle: $\dot\gamma_w = \dfrac{3n + 1}{4n} \left(\dfrac{4Q}{\pi R^3}\right)$ and $\tau_w = \dfrac{R\Delta P}{2L_d}$ (4.78)

slit: $\dot\gamma_w = \dfrac{2n + 1}{3n} \left(\dfrac{3Q}{2C^2w}\right)$ and $\tau_w = \dfrac{C\Delta P}{L_d}$ (4.79)

where n is the flow behavior index. Once $\dot\gamma_w$ is known, η can be determined from data or a rheology model for the food dough being considered. Since η is the ratio of shear stress to shear rate, and both are known from above, the pressure drop across the die can be calculated directly.

B. End Effects

The use of an equation such as Equation 4.27 assumes that there is no pressure drop associated with end effects on the die. It is well known from fluid mechanics that significant pressure drops occur at the constriction at the entrance of the die and the expansion at the exit. To reduce entrance losses, it is common to taper the entrance on a die insert as shown in Figure 24 which also illustrates the streamlines of flow (lines of constant velocity) through the die.

The end correction for an individual die hole can be considered as in Chapter 3 where

$$\tau_w = \frac{\Delta P}{2\left(\dfrac{L}{R} + \dfrac{L^*}{R}\right)} \tag{3.47}$$

where τ_w = effective shear stress at wall and L^*/R = equivalent length to radius ratio for a hole of zero length. In this formulation, the end correction is the pressure drop caused by an equivalent length of die having radius, R. For dies with a tapered entrance, L^*/R would be approximately five. The specific value of L^*/R can only be found by experimentation.

C. Irregular Shapes

Many food dies have an irregular shape. As an approximation, the hydraulic radius of the irregular die can be used in the equations for circular dies. Hydraulic radius has been defined as

$$R_H = s/\psi \tag{4.80}$$

where R_H = hydraulic radius, s = cross-sectional area of die, and ψ = wetted perimeter of die.

D. Design

Die design on food extruders consists of selecting the length and number of dies to achieve the required product at a given rate, Q. An operating pressure for the extruder and maximum shear stress in the die must also be known. The maximum τ is that which will give a smooth surface in the extruded product with no noticeable feathering or surface roughness and can be considered to be $<7\times10^5$ N/m^2.

In die design, $\dot{\gamma}_w$ is first calculated that gives the maximum τ_w from rheology data at the extrusion conditions for the food. The flow through a single die can be calculated from $\dot{\gamma}_w$ using Equation 4.78. The number of die holes is the total flow divided by the flow through each hole. The length of the die hole is next calculated using Equation 3.47 considering end effects.

1. Example 5

A cooked cereal dough at 90°C and 28% moisture has a viscosity given by $\eta = 8700(\dot{\gamma})^{-0.49}$, Ns/m^2. Dough density can be assumed to be 1250 kg/m^3. The die will produce rod-shaped pieces 0.5 cm in diameter at a flow rate of 1 MT/hr. For purposes of design, keep the shear stress at the wall of the die $<3\times10^5$ N/m^2 and assume the extruder will develop a back pressure of 3.5 MPa (34 atm). Assuming the die holes exhibit no end effects, estimate the number of die holes and the length of the die hole. Assume $L^*/R = 3$, to account for end effects, recalculate the length of the die hole. For the solution:

1. Calculate volumetric flow, Q

$$Q = \frac{\dot{m}}{\rho} = \frac{2000\ kg}{h} \left| \frac{m^3}{1250\ kg} \right| \frac{1hr}{3600\ s} \left| \frac{10^6\ cm^3}{m^3} \right. = 444\ cm^3/s$$

Calculate flow through each die hole, Q_{di} so that maximum τ_w is not exceeded. Since

$$\eta = 8700(\dot{\gamma})^{-0.49}$$

then

$$\tau = 8700(\dot{\gamma})^{0.51}, \text{ N/m}^2$$

Solving for $\dot{\gamma}_w$ corresponding to $\tau_w = 3\times10^5$ N/m^2

$$3.0 \times 10^5 = 8700(\dot{\gamma}_w)^{0.51}$$

Since

$$\dot{\gamma}_w = 1035 \text{ s}^{-1}$$

$$\dot{\gamma}_w = \frac{3n + 1}{4n} \left(\frac{4Q_{di}}{\pi R^3} \right) \qquad (4.78)$$

$$Q_{di} = 1035 \left[\frac{4(0.51)}{3(0.51) + 1} \right] \frac{\pi(0.3)^3}{4} = 17.7 \text{ cm}^3/\text{s}$$

Number of dies are calculated as

$$\# = Q/Q_{di} = 444/17.7 = 25.1$$

Use 25 dies. Calculate length of die hole.

$$\tau_w = \Delta PR/2L$$

or

$$L = \frac{\Delta PR}{2\tau_w} = \frac{3.5 \times 10^6 \text{ N}}{\text{m}^2} \left| \frac{0.3 \text{ cm}}{2} \right| \frac{\text{m}^2}{3 \times 10^5 \text{ N}} = 1.75 \text{ cm}$$

For the die, $L/R = 1.75/0.3 = 5.8$.
  2. Recalculate length of die using

$$\tau_w = \frac{\Delta P}{2\left(\dfrac{L}{R} + \dfrac{L^*}{R}\right)} \qquad (3.47)$$

Then,

$$L = R\left[\frac{\Delta P}{2\tau_w} - \frac{L^*}{R}\right] = 0.3 \left[\frac{3.5 \times 10^6 \text{N}}{\text{m}^2} \left| \frac{}{2} \right| \frac{\text{m}^2}{3 \times 10^5 \text{ N}} - 3 \right] = 0.85 \text{ cm}$$

For the die, $L/R = 0.85/0.3 = 2.83$. Without considering end effects, die would have been 106% too long.

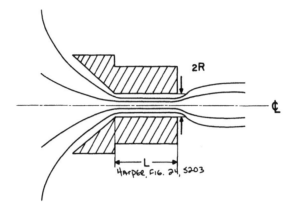

FIGURE 24. Cross section of die insert illustrating streamlines (lines of constant velocity) of flow.

E. Moisture Loss

An energy balance can be written around the discharge die to estimate the loss of moisture in the food dough caused by the flashing of steam as the pressure is reduced to ambient. This balance shows

$$Q\rho c_p (T_1 - T_2) = Q\rho(M_1 - M_2)\Delta H_{fg} \qquad (4.81)$$

or

$$M_2 = \frac{M_1 \Delta H_{fg} - c_p(T_1 - T_2)}{\Delta H_{fg}} \qquad (4.82)$$

where M = moisture fraction wet basis, ΔH_{fg} = latent heat of vaporization at ambient pressure, c_p = specific heat of food, subscripts 1 = before product emerges from die, and 2 = after product emerges from die. The development of Equation 4.82 assumes that the loss of moisture is small compared to $Q\rho$, so the mass flow of material before and after the die is nearly identical, flashing of water occurs at the normal boiling point, and the specific heat changes little with pressure or temperature.

1. Example 6

For a food product which has a specific heat, c_p = 2.7 kJ/kg°C, calculate the moisture loss experienced when the product cools from 170 to 70°C when the initial moisture is 30%.
Solution:

$$\Delta H_{fg} = 2675.8 \text{ kJ/kg at } 100°\text{C}$$

Using Equation 4.82

$$M_2 = \frac{0.30(2675.8) - 2.7(170 - 70)}{2675.8} = 0.199 \text{ or } 19.9\%$$

VI. SCALE-UP

Scale-up is a problem faced by the food engineer who must take a pilot plant developed extrusion process and design a full scale production plant. Examination of Equations 4.22 and 4.66 for the metering section for flow and power with the assumptions stated provides some important information which is useful in extruder scale-up, which is very nonlinear. If pressure flow in Equation 4.22 is small compared to drag flow, volumetric flow would be expected to increase in relation to Equation 4.23 or $G_1 = f(D^2H)$ for constant N. If all dimensions in the extruder were increased by a scale factor ϕ, then flow would theoretically increase by ϕ^3.

Similarly, input power should increase in relation to Equation 4.67 where $G_8 = f(D^2L)$ or ϕ^3 also. The cubic scale-up of both flow and power would indicate that for a doubling of extruder size, output and energy requirements should increase by a factor of 2^3 or 8.

In most food extrusion applications, heat transfer through jackets around the barrel and the developed heat produced by viscous dissipation of mechanical energy play a significant role in the operation of the extruder. Both processes are surface area related. An appropriate equation for heat transfer between the heat transfer media in jackets and the food product would be

$$q = UA\Delta T \tag{4.83}$$

where q = rate of heat transfer, U = overall heat transfer coefficient or

$$U = \frac{1}{\dfrac{1}{h_i} + \dfrac{\Delta x}{k} + \dfrac{1}{h_o}} \tag{4.84}$$

h_i, h_o = inside and outside surface heat transfer coefficients, respectively, k = thermal conductivity of jacket wall, Δx = thickness of jacket wall, A = heat transfer surface area = πDL, and ΔT = temperature difference. The overall heat transfer coefficient U will vary with h_i, the one coefficient which will change with scale-up. If the screw flights passing over the surface of the barrel simulate a scraped surface heat exchanger, h_i would be a function of $(DN)^m$, where m is less than 1.0. Therefore, as D becomes larger, h_i would increase slightly, thus increasing U. The small increase in U is insignificant, however, compared to the changes in surface area, A, which increases as ϕ^2. For proper scale-up of extruder operations, it is necessary to adjust D, H, and N in such a manner that the heat transfer area, flow rate, and mechanical energy input increase at approximately the same rate.

Another factor affecting scale-up is the shear rate in the channel of the screw and the clearance over the land of the flight. Because most food materials are non-Newtonian, it is important that these shear rates remain the same for both extruders. Shear rate in the screw channel is given by,

$$\dot{\gamma}_H = f\left(\frac{V}{H}\right) \approx \frac{\pi DN}{H} \tag{4.41}$$

Shear rate in the clearance, δ, can be written

$$\dot{\gamma}_\delta = \left(\frac{V}{\delta}\right) = \frac{\pi DN}{\delta} \tag{4.85}$$

where $\dot{\gamma}_o$ = shear rate in the clearance, s^{-1}.

Residence time for the food in the extruder is another important parameter which should remain constant, if possible, during scale-up. A first order approximation of the average residence time can be made by dividing the volume of the screw by the drag flow as

$$\bar{t} = \frac{V_o}{Q} = f\left(\frac{DHL}{D^2HN}\right) = f\left(\frac{L}{DN}\right) \tag{4.86}$$

where t = average residence time, V_o = volume of scew channel, and Q = flow $\simeq$ G_1N.

Using these relationships, several possible scale-up techniques are shown in Table 1. For most scale-ups, the dimensional alterations described in Case 3 of Table 1 will result in the extruder producing product which most closely resembles the prototype made on the smaller extruder. Notice that the screw flight height and speed of rotation are altered to reduce the theoretical increase in output resulting from a purely geometric scale-up. Most importantly, the apparent shear rates in the channel and clearance are held constant. The residence time for the food, however, is longer. This increase in residence time coupled with increased mechanical energy dissipated in the channel of $\sim\phi^{2.5}$ results in the necessity to do some additional barrel/screw cooling.

Since a large number of simplifying assumptions were made in deriving these theoretical relationships, some deviations from precise scale-up should be expected. Normally, these can be compensated for through small changes in jacket temperatures, screw speeds, or feed moistures. The scale-up proposed in Case 3 nearly maintains the relative inputs of viscous dissipation energy and heat transfer from the jackets to the product flow that were achieved in the original extruder.

The die also needs to be considered in scale-up. The number of die holes should be increased by the anticipated scale-up factor. If the scale-up were achieved as proposed in Case 3, the number of die holes of the same size used on the small extruder would be increased by ϕ^2. In this manner the pressure drop across the die would be similar for both the large and small extruder because of the increased flow area available is proportional to the anticipated increased flow. Also, the shear rate of the dough passing through the die would remain constant.

A. Example 7

In laboratory tests, a small extruder was used to produce a new food product at the rate of 50 kg/hr. A large extruder is to be specified which has a rated output of 500 kg/hr. If the pilot scale extruder had the dimensions shown, recommend the appropriate dimensions for the full scale extruder and the speed and size of the drive.

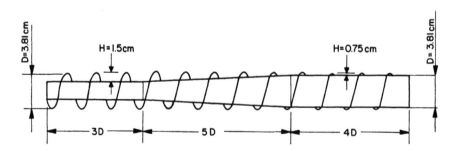

Table 1
SCALE-UP ALTERNATIVES FOR FOOD EXTRUDERS USING SCALE-UP FACTOR ϕ [a]

	Dimension					Output variables						Comments
Case	D	H	δ	L	N	Q	E	$\dot{\gamma}_n$	$\dot{\gamma}_s$	q	$\bar{\tau}$	
1	ϕ	ϕ	ϕ	ϕ	1	ϕ^3	ϕ^3	1	1	$\sim\phi^2$	1	Larger extruders will be limited by area for shear and heat transfer
2	ϕ	ϕ	ϕ	ϕ	$1/\phi$	ϕ^2	ϕ^2	$1/\phi$	$1/\phi$	ϕ^2	ϕ	Mechanical energy input would be greatly reduced along with thermal uniformity
3	ϕ	$\sqrt{\phi}$	$\sqrt{\phi}$	ϕ	$1/\sqrt{\phi}$	ϕ^2	$\sim\phi^{2.5}$ [b]	1	1	$\sim\phi^2$	$\sqrt{\phi}$	Mechanical energy input to channel and residence time will be increased

[a] Geometric scale-up factor, ϕ.

[b] Assumes μ W/H (cos^2 θ + 4sin^2 θ), energy dissipated in channel, is much less than the energy dissipated in clearance, μ_e e/δ.

From Harper, J. M., *Food Technol. (Chicago)*, 32(7), 67, 1978. Copyright by Institute of Food Technologists. With permission.

p = 1

D = 3.81 cm (straight)

H = 0.75 cm (metering)

H = 1.50 cm (feed)

δ = 0.1 cm

N = 80 rpm

L/D = 12:1

$E/\dot{m}$ = 0.03 kWhr/kg

θ = 17.6°

DIE = 1 hole, 0.5 cm dia. × 0.7 cm long

$\bar{t}$ = 125 sec

Solution — Scale-up will be attempted using the procedure outlined by Case 3 in Table 1.

$$\phi^2 = \frac{Q_2}{Q_1} = \frac{500}{50} = 10$$

$$\phi = \sqrt{10} = 3.16$$

$$\sqrt{\phi} = \sqrt{3.16} = 1.78$$

Dimensions should be altered as follows:

$D_2 = D_1\phi = (3.81)(3.16) = 12.0$ cm

H_2 metering $= H_1\sqrt{\phi} = 0.7(1.78) = 1.25$ cm

H_2 feed $= H_1\sqrt{\phi} = 1.5(1.78) = 2.67$ cm

$\delta_2 = \delta_1\sqrt{\phi} = 0.1(1.78) = 0.178$ cm

$L_2 = L_1\phi = 12(3.81)(3.16) = 144.5$ cm

L_2 metering $= 4(3.81)(3.16) = $ 48.2 cm

L_2 trans $= 5(3.81)(3.16) = $ 60.2 cm

L_2 feed $= 3(3.81)(3.16) = $ $\underline{36.1 \text{ cm}}$

144.5 cm

$$N_2 = \frac{1}{\sqrt{\phi}} N_1 = \frac{80}{1.78} = 45 \text{ rpm}$$

Die #2 = #1ϕ^2 = 10 die holes, keeping die size the same. The expected result would be

$$E_2\dot{m}_2 = E_1\dot{m}_1\phi^{2.5} = (0.03)(50)(1.78) = 26.7 \text{ kW}$$

$$\bar{t}_2 = \bar{t}_1 \sqrt{\phi} = 125(1.78) = 222 \text{ sec}$$

The extrusion scale-up it is often very difficult to achieve precisely the same product produced on smaller equipment. In the case examined, mechanical energy input will be larger causing a potentially greater temperature rise in the product. The higher cooking temperature may be partially offset by cooling the jackets surrounding the barrel.

REFERENCES

1. ASTM, Static and kinetic coefficients of friction of plastic film and sheeting, in *Annual Book of ASTM Standards*, American Society for Testing and Materials, Philadelphia, Pa., 35, 1978, 595.
2. Booy, M. L., Influence of channel curvature on flow, pressure distribution and power requirements of screw pumps and melt extruders, *SPE Trans.*, 3(3), 176, 1963.
3. Booy, M. L., Influence of oblique channel ends on screw-pump performance, *Polym. Eng. Sci.*, 7(1), 5, 1967.
4. Bruin, S., van Zuilichem, D. J., and Stolp, W., Fundamental and engineering aspects of extrusion of biopolymers in a single-screw extruder, *J. Food Proc. Eng.*, 2(1), 1, 1978.
5. Carley, J. F., Problems of flow in extrusion dies, *SPE J.*, 19, 1263, 1963.
6. Carley, J. F., Single-screw pumps for polymer melts, *Chem. Eng. Prog.*, 58(1), 53, 1962.
7. Carley, J. F. and Strub, R. A., Basic concepts of extrusion, *Ind. Eng. Chem.*, 45, 970, 1953.
8. Carley, J. F., Mallouk, R. S., and McKelvey, J. M., Simplified flow theory for screw extruders, *Ind. Eng. Chem.*, 45, 974, 1953.
9. Colwell, R. E. and Nickolls, K. R., The screw extruder, *Ind. Eng. Chem.*, 51(7), 841, 1959.
10. Darnell, W. H. and Mol, E. A. J., Solids conveying in extruders, *SPE J.*, 12, 20, 1956.
11. Fenner, R. T., Designing extruder screws and dies with the aid of computers, in *New Technology in Extrusion and Injection Molding*, Plastic Institute, London, 1973.
12. Fenner, R. T., *Extruder Screw Design*, Iliffe Books, London, 1970.
13. Fricke, A. L., Clark, J. P., and Mason, T. F., Cooking and drying of fortified cereal foods: extruder design, *Chem. Eng. Prog. Symp. Ser.*, 73(163), 134, 1977.
14. Griffith, R. M., Fully developed flow in screw extruders, *Ind. Eng. Chem. Fundam.*, 1(3), 180, 1962.
15. Harmann, D. V. and Harper, J. M., Effects of extruder geometry on torque and flow, *Trans. ASAE*, 16(6), 1175, 1973.
16. Harmann, D. V. and Harper, J. M., Modeling a forming foods extruder, *J. Food Sci.*, 39, 1099, 1974.
17. Harper, J. M., Extrusion processing of food, *Food Technol. (Chicago)*, 32(7), 67, 1978.
18. Harper, J. M., Goals and activities of the Colorado State University LEC program, in *Low-Cost Extrusion Cookers*, LEC-1, Harper, J. M. and Jansen, G. R., Eds., Colorado State University, Fort Collins, 1976, 19.
19. Harper, J. M., Rhodes, T. P., and Wanninger, L. A., Jr., Viscosity model for cooked cereal doughs, *Chem. Eng. Prog. Symp. Ser.*, 67(108), 40, 1971.
20. Jasberg, B. K., Mustakas, G. C., and Bagley, E. B., Extrusion of Defatted Soy Flakes — Model of a Plug Flow Process, *J. Rhea.*, 23(4), 437, 1979.
21. Lancaster, E. B., Specific volume and flow of corn grits under pressure, *Chem. Eng. Prog. Symp. Ser.*, 67(108), 30, 1971.
22. McKelvey, J. M., *Polymer Processing*, John Wiley & Sons, New York, 1962.
23. Middleman, S., *Fundamentals of Polymer Processing*, McGraw-Hill, New York, 1977.
24. Mustakas, G. C., Albrecht, W. J., Bookwalter, G. N., McGhee, J. E., Kwolek, W. F., and Griffin, E. L., Extruder-processing to improve nutritional quality, flavor and keeping quality of full-fat soy flour, *Food Technol. (Chicago)*, 24, 1290, 1970.
25. Paton, J. B., Squires, P. H., Darnell, W. H., Cash, F. M., and Carley, J. F., Extrusion, in *Processing of Thermoplastic Materials*, Bernhardt, E. C., Ed., Robert E. Krieger, Huntington, N.Y., 1974.
26. Rowell, H. S. and Finlayson, D., Screw viscosity pumps, *Engineer*, 114, 606, 1922.
27. Rowell, H. S. and Finlayson, D., Screw viscosity pumps, *Engineer*, 126, 249, 1928.
28. Sahagun, J. and Harper, J. M., Parameters affecting the performance of a low-cost extrusion cooker, *J. Food Proc. Eng.*, 3(4), 199, 1980.
29. Squires, P. H., Screw extruder pumping efficiency, *SPE J.*, 14(5), 24, 1958.
30. Squires, P. H. and Galt, J. C., unpublished data, 1958.
31. Tadmor, Z., Fundamentals of plasticating extrusion, *Polym. Eng. Sci.*, 6, 185, 1966.
32. Tadmor, Z. and Klein, I., *Engineering Principles of Plasticating Extrusion*, Van Nostrand Reinhold, New York, 1970.
33. Tsao, T. F., Harper, J. M., and Repholz, K. M., The effects of screw geometry on extruder operational characteristics, *AIChE Symp. Ser.*, 74(172), 142, 1978.

Chapter 5

RESIDENCE TIME DISTRIBUTION AND STRAIN

I. INTRODUCTION

The cooking of food in an extruder has been previously described as a series of chemical and physical changes occurring because of heating and shearing during the time of passage through the extruder. Cooking encompasses protein denaturation and cross-linking, starch gelatinization and dextrinization, browning, denaturization of vitamins and enzymes, etc., and is obviously very complex. The resulting properties of the cooked food are a composite or weighted average of all components which emerge from the extruder die. Ideally, all food particles would receive the same time-temperature-shear history in their passage through the screw channel, but this is not the case due to the varying velocities within different parts of the screw channel.

An examination of the factors which effect the residence time distributions (RTDs) for isothermal Newtonian flow is helpful in understanding the effects of extrusion variables and the screw geometry. The RTD will be developed from the velocity profiles. From the RTD, it is possible to describe mixing within the extruder using the average total strain received by the food during its passage. Consequently, examination of experimental techniques to measure RTDs for actual extrusion processes will be described and discussed.

II. RESIDENCE TIME

A. Channel Flow

The time an individual particle of Newtonian fluid spends in the metering section of an extruder can be calculated from the velocity distribution as originally done by McKelvey.[5] These velocity distributions within the channel are given by Equations 4.15 and 4.20 and are

$$v_z = V_z[(1 - 3a)(y/h) + 3a(y/H)^2] \tag{5.1}$$

and

$$v_x = V_x[2 - 3(y/H)]\frac{y}{H} \tag{5.2}$$

where $V_z = \pi DN \cos \theta$, $V_x = -\pi DN \sin \theta$, $a = -Q_p/Q_d$ (Equation 4.11), and $y/H =$ position within channel. The velocity distribution equations clearly show that a particle of food will move in a helical path down the channel of the screw as shown in Figure 1. When the position of a particle is $y > 2/3 H$, it will move in a $-x$ direction until it reaches the leading edge of the next flight. At this point, the particle will move to a complementary position $y_c < 2/3 H$ and start transversing the channel in a $+x$ direction. Once it reaches the trailing edge of the leading flight, it will return to its corresponding y position in the upper third of the channel. There is a unique relationship between y/H and y_c/H for particles circulating within the channel as shown in Figure 2 and given by

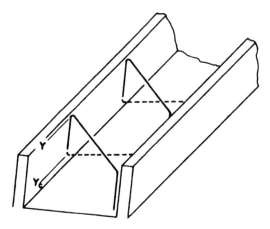

FIGURE 1.   Circulation of a fluid particle as it moves
through screw channel. (Reprinted with permission
from Bigg, D. and Middleman, S., *Ind. Eng. Chem.
Fundam.*, 13(1), 66, 1974. Copyright © 1974 American
Chemical Society.)

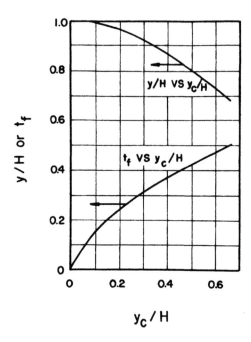

FIGURE 2.   Particle position y/H (upper third
of channel) vs. its position y_c/H (lower two
thirds of channel). Also t_f vs. y_c/H. (From
McKelvey, J. M., *Polymer Processing*, John
Wiley & Sons, New York, 1962. Copyright ©
1962 John Wiley & Sons. Reprinted by permission of John Wiley & Sons.)

$$\frac{y}{H} = \left(\frac{1 - \frac{y_c}{H}}{2}\right)\left[1 + \left(\frac{1 + 3\frac{y_c}{H}}{1 - \frac{y_c}{H}}\right)^{\frac{1}{2}}\right] \tag{5.3}$$

The above description of the flow within the channel assumes that the presence of the flights does not disturb the flow pattern other than creating the circulating flow pattern. Since there is no net flow in the x direction, then

$$\int_{0}^{y_c} v_x dy = -\int_{y}^{H} v_x dy \tag{5.4}$$

As described above, v_x changes direction below a plane at $y/H = 2/3$.

The fraction of time that a particle spends in the upper third of the channel is a function of its relative position in the lower portion of the channel. Using the relationship between y/H and y_c/H in Equation 5.3, the particle velocity in the upper and lower portions of the channel for any corresponding pair of positions can be calculated using Equation 5.2. The time necessary to transverse the channel width, W, in the x direction is

$$t_u = \frac{W}{v_x(y)} \tag{5.5}$$

and

$$t_\ell = \frac{W}{v_x(y_c)} \tag{5.6}$$

where $v_x(y)$ and $v_x(y_c)$ are the velocities at y and y_c, respectively, and t_u and t_ℓ are the times spent in the upper and lower portions of the channel, respectively. The fraction of time, t_f, spent in the upper portion is

$$t_f = \frac{t_u}{t_u + t_\ell} = \frac{1}{1 - \frac{\frac{y}{H}\left(2 - 3\frac{y}{H}\right)}{\frac{y_c}{H}\left(2 - 3\frac{y_c}{H}\right)}} \tag{5.7}$$

This functional relationship between t_f and y_c/H is also shown in Figure 2. For a particle that moves near the root of the channel, very little of its total time is spent in the upper third of the channel.

It is now clear that depending upon the initial position of a particle within the screw channel, its residence time will vary. The residence time for a particle at y, $t(y)$, is defined as

$$t(y) = \frac{L}{\bar{v}_\lambda(y)} \tag{5.8}$$

where L is the axial length of the screw and $\bar{v}_\lambda$ is the average particle velocity in the axial direction. The vectorial addition of v_x and v_z gives

$$\frac{\bar{v}_\lambda}{V} = 3\frac{y}{H}\left(1 - \frac{y}{H}\right)(1 - a)\cos\theta\sin\theta \qquad (5.9)$$

where $V = \pi DN$. Since $\bar{v}_\lambda$ is a function of the position of the particle and the fraction of time spent in the upper and lower portions of the channel, then

$$\bar{v}_\lambda(y) = v_\lambda(y)t_f + v_\lambda(y_c)(1 - t_f) \qquad (5.10)$$

Combining Equations 5.9 and 5.10 gives residence time as a function of a particles position in the channel or

$$t(y) = \frac{L}{3V(1 - a)\left[\frac{y_c}{H}\left(1 - \frac{y_c}{H}\right) + t_f\left(\frac{y}{H} - \frac{y_c}{H}\right)\left(1 - \frac{y}{H} - \frac{y_c}{H}\right)\right]\sin\theta\cos\theta} \qquad (5.11)$$

Evaluation of Equation 5.11 requires the understanding of the relationship between y/H and y_c/H as given in Equation 5.3.

A dimensionless residence time can be obtained by dividing Equation 5.11 by $L/3V(1-a)\sin\theta\cos\theta$. Examination shows that the actual residence time in an extruder will increase by making L longer, reducing V by decreasing N or increasing a. These statements are in agreement with the residence time considerations discussed in Chapter 4 where it was concluded $t(y) = f(N^{-1})$.

The relationship between $t(y)$ and the bulk average residence time

$$\bar{t} = \frac{pHWL}{Q\sin\theta} \qquad (5.12)$$

as defined by the ratio of the channel volume and volumetric flow rate, is given in Figure 3. Particles which have a position $y/H = 2/3$ have a minimum residence time, t_0, while particles at either the channel root or barrel surface have theoretically infinite residence times.

B. RTD-Newtonian Flow

The RTD is a plot of residence time against the fraction of flow having this residence time. These distributions have been best analyzed in terms of E- and F-diagrams commonly used to describe the exit time response of a system to a pulse perturbation in concentration of a dough constituent at the input of the system. Levenspiel[4] provides additional information on these diagrams.

Figure 4 describes both the E(t)- and F(t)-functions. When an instantaneous pulse is injected into a system, the output will show a skewed distribution of this change called the "residence time distribution function". The fraction of flow that has spent between t and t + dt is called the E(t) function.

The relationship between E(t) and F(t) is given as

$$F(t) = \int_0^t E(t)dt \qquad (5.13)$$

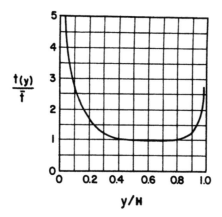

FIGURE 3. Relative residence time as a function of the initial particle position. (From McKelvey, J. M., *Polymer Processing*, John Wiley & Sons, New York, 1962, 322. Copyright © 1962 John Wiley & Sons. Reprinted by permission of John Wiley & Sons.)

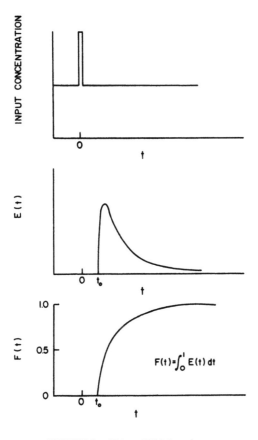

FIGURE 4. E(t) and F(t) functions.

Clearly, F(t) is the cumulative distribution of E(t) which shows a sharp rise starting at the minimum residence time, t_o, and equals 1.0 at $t = \infty$.

Pinto and Tadmor[6] have analyzed single screw extruder RTDs for Newtonian fluids with constant μ. Their analysis shows

$$E(t)dt = \frac{3\frac{y}{H}\left\{1 - \frac{y}{H} + \left[1 + 2\frac{y}{H} - 3\left(\frac{y}{H}\right)^2\right]^{1/2}\right\}}{\left[1 + 2\frac{y}{H} - 3\left(\frac{y}{H}\right)^2\right]^{1/2}} \qquad (5.14)$$

and a function of y/H. Equation 5.14 can be evaluated when Equation 5.3, the relationship between y/H and y_c/H, is applied. The average residence time, $\bar{t}$, is related to E(t) by

$$\bar{t} = \int_0^\infty tE(t)dt \qquad (5.15)$$

In a single screw extruder, $\bar{t} = 4/3\ t_o$ for Newtonian fluids.

Figure 5 shows an E(t) diagram for a Newtonian fluid against the dimensionless time variable, $t/\bar{t}$. The E(t) for an extruder lies somewhere between that for laminar flow in a pipe and plug flow, where no back mixing exists.

It is possible to show by integrating Equation 5.13, using the value of E(t)dt in Equation 5.14, that

$$F(t) = \frac{1}{2}\left\{3\left(\frac{y}{H}\right)^2 - 1 + \left(\frac{y}{H} - 1\right)\left[1 + \frac{y}{H} - 3\left(\frac{y}{H}\right)^2\right]^{1/2}\right\} \qquad (5.16)$$

Again, F(t) is related to residence time by Equations 5.3 and 5.11 and varies with particle position y. The F(t) function for Newtonian fluids is shown in Figure 5.6 as a function of $t/\bar{t}$. Completely backed mixed flow exists when total mixing occurs within the flow volume. The F(t) is in a generalized form; therefore, it is suitable for use with extrusion of Newtonian fluids.

C. RTD-Non-Newtonian Flow

Bigg and Middleman[1] have extended the work of Pinto and Tadmor[6] on RTDs to include isothermal non-Newtonian flow. Instead of using the velocity profiles for a Newtonian fluid given in Equations 5.1 and 5.2, they used the velocity profiles existing for a power law fluid similar to Griffith[3] who developed the equations for flow for power law fluids. For a power law fluid, F(t) is both a function of $t/\bar{t}$ and the dimensionless parameter defined by Equation 4.47.

$$G_z = \frac{P_2 - P_1}{L} \cdot \frac{H^{n+1} \sin\theta}{m(\pi DN)^n} \qquad (5.17)$$

where $P_2 - P_1/L$ = pressure rise long screw, H = flight height, m = consistency index in Equation 3.7, n = exponent in power law, D = diameter, and N = speed.

Figure 7 shows F(t) for a non-Newtonian fluid having n = 0.4 and various values of $G_z/\cos\theta$. It can be seen that as $G_z/\cos\theta$ increases, F(t) approaches the value which would be characteristic of plug flow. As n is reduced to 0.2, Bigg and Middleman[1] showed the F(t) curves approached those which result with complete back mixing.

The delay time for the appearance of an impulse change in concentration at the feed of an extruder is a function of n and the reduced dimensionless flow parameter, Q/

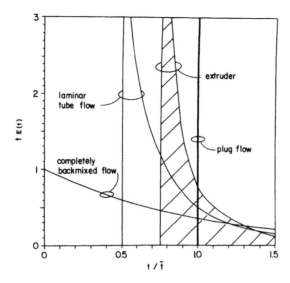

FIGURE 5. E(t) for Newtonian flow in an extruder. (From Bruin, S., van Zuilichem, D. J., and Stolp, W., *J. Food Proc. Eng.*, 2(1), 1, 1978. With permission. © Food and Nutrition Press, Inc.)

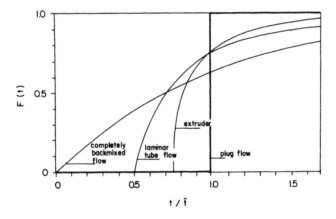

FIGURE 6. F(t) for Newtonian flow in an extruder. (From Bruin, S., Van Zuilichem, D. J., and Stolp, W., *J. Food Proc. Eng.*, 2(1), 1, 1978. With permission. © Food and Nutrition Press, Inc.)

$2Q_d$. In this parameter, Q is the measured volumetric discharge and Q_d is the theoretical drag flow (assuming all correction factors are equal to unity and no leakage flow exists), which is given by G_1N (Equation 4.23). The dimensionless delay time, $t_o/\bar{t}$, is shown in Figure 8. For a non-Newtonian fluid, the delay time can be faster or slower than $t_o = 3/4t$ for Newtonian fluids; the value depending on operating conditions. Figure 8 suggests a quick way of estimating n by measuring the delay time and flow rate.

D. Measurement of RTDs

The measurement and reporting of residence time distributions in food extrusion is very important. van Zuilichem et al.[8] reported measuring RTDs in corn grits extrusion,

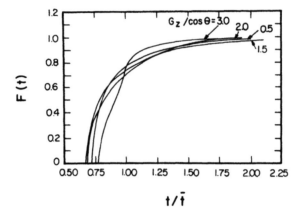

FIGURE 7. F(t) curves for n = 0.4 at various values of $G_z/\cos\theta$. (From Bigg, D. and Middleman, S., *Ind. Eng. Chem. Fundam.*, 13(1), 66, 1974. With permission. Copyright © 1974 American Chemical Society.)

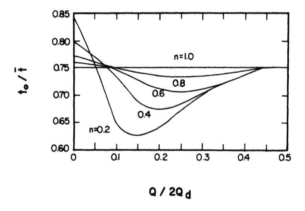

FIGURE 8. Normalized delay time as a function of dimensionless flow. (From Bigg, D. and Middleman, S., *Ind. Eng. Chem. Fundam.*, 13(1), 66, 1974. With permission. Copyright © 1974 American Chemical Society.)

using a radio-tracer technique with ^{64}Cu, where E(t) was obtained with a scintillation counter. ^{64}Cu was chosen as a tracer because it has a half-life of 12.7 hr. Despite this relatively short half-life, the cost of the sources, counters, and extensive safety requirements associated with radio-active sources makes practical application of the techniques difficult.

Residence time data were obtained by Bigg and Middleman[1] by using a red food coloring dissolved in the liquids being extruded. When the extruder was operating steadily and the first flights of the feed section were exposed, the dyed sample was introduced to the feed hopper and a timer started. Samples of extrudate were collected and the light transmittance of the sample was used to detect the emergence of the red color.

Tsao[7] used a similar dye technique to measure residence times in a dry cereal extrusion. Five milligrams of a fluorescent dye, rhodamine B, was added to the exposed screw through the feed hopper. At the extruder discharge, portions of the puffed ex-

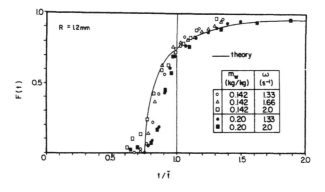

FIGURE 9. F(t) for extrusion of corn grits in a single screw extruder. (From Bruin, S., van Zuilichem, D. J., and Stolp, W. J., *J. Food Proc. Eng.*, 2(1), 1, 1978. With permission. © Food and Nutrition Press, Inc.)

trudate were cut at 5-s intervals. Each of these was blended with 500 cm³ of water and centrifuged. The dye concentration in the supernatent was measured with a Perkin-Elmer MPF 2A fluorescence spectrophotometer at an excitation wavelength of 492.5 mµ and an emission wavelength of 585 mµ. Using a standard curve previously constructed, a plot of dye concentration against time gave measurement of E(t).

Once a measure of E(t) is obtained, the normalized F(t) can be calculated by the numerical integration of E(t)dt and plotted against t/t̄ where t is defined by Equation 5.12. In food extrusion, most biopolymeric materials exhibit non-Newtonian behavior and so it is necessary to report η and $G_z/\cos\theta$ along with F(t) if a comparison is to be made with other food materials or extrusion conditions.

Some deviation of measured F(t) diagrams with those shown in Figures 6 and 7 should be expected. In most food extruders, the screw geometry is not constant and, therefore, changes in flow patterns in the channel markedly affect E(t) and F(t). Ideally, the reduced F(t) adequately characterizes the RTD for food extruders (as discussed in the next section), but deviations from simplified flow and the assumptions used to derive these distributions exist so some actual nonideality should be expected.

Finally, van Zuilichem et al.[8] and Tsao[7] both described the food extruder as a region for plug flow followed by two perfect mixers in series. Such a description allowed them to easily model the E(t) diagram and smooth the scatter in the experimental data they took.

E. RTDs in Food Extrusion

van Zuilichem et al.[8] were the first to measure RTDs in the extrusion of biopolymers. They extruded corn grits at 14 and 20% moisture content and at three screw speeds with a plastics type extruder having a 48-mm diameter compressive screw having a compression ratio of 2.5:1. Their data are plotted as an F(t) diagram in Figure 9. The deviations of the data with the F(t) curve for Newtonian flow can be rationalized in terms of non-Newtonian effects previously discussed. It is encouraging to see that the F(t) distribution for the isothermal extrusion of a Newtonian fluid so nearly fits the data for a strongly non-Newtonian material (n ≈ 0.5) which underwent cooking through the extrusion process and consequently had significant viscosity changes in different portions of the screw.

Bigg and Middleman[1] measured F(t) for Newtonian corn syrups and non-Newtonian resin solutions with a 1.90 cm Brabender® extruder. In all cases, they found excellent

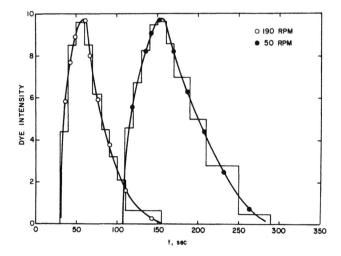

FIGURE 10.   Relative dye intensity vs. residence time.

agreement between the experimental results and their theory which was presented earlier.

1. Example 1

A laboratory Brabender® extruder 1.90 cm in diameter having a 20:1 L/D ratio and a screw with a 1:1 CR was used to extrude a defatted soy protein dough containing approximately 30% water. Such doughs are very non-Newtonian and the flow behavior index, n, has been reported to be about 0.2.

A dye was injected into the dough as an impulse and the relative color intensity was measured in the extrudate as a function of time. The resulting E(t) diagrams for two different extrusion speeds are shown in Figure 10. Calculate the F(t) for these two cases and compare the results with the F(t) for a Newtonian fluid.

For data at 190 rpm:

t	$t/\bar{t}$	Area int × time	E(t)	F(t)
35	0.78	4.4 × 10 = 44	0.088	0.088
45	1.00	8.5 × 10 = 85	0.171	0.259
55	1.22	9.6 × 10 = 96	0.193	0.452
65	1.44	8.5 × 10 = 85	0.171	0.623
75	1.67	6.2 × 10 = 62	0.124	0.747
85	1.89	4.5 × 10 = 45	0.090	0.837
95	2.11	3.2 × 10 = 32	0.064	0.901
105	2.33	2.1 × 10 = 21	0.042	0.943
132	2.93	0.6 × 45 = __27__	0.054	0.997
		497		

Assume Newtonian behavior to estimate $\bar{t}$ from data.

$$\bar{t} = (4/3)t_o = (4/3)(34) = 45 \text{ s}$$

This assumption will lead to some error in the F(t) curve. A better method would be to use Figure 8 but this requires knowing $Q/2Q_d$.

For data at 50 rpm:

t	$t/\bar{t}$	Area int × time	E(t)	F(t)
115	0.80	4.6 × 10 = 46	0.054	0.054
125	0.87	6.7 × 10 = 67	0.079	0.133
135	0.94	8.3 × 10 = 83	0.098	0.231
145	1.01	9.5 × 10 = 95	0.112	0.343
155	1.08	9.6 × 10 = 96	0.114	0.457
165	1.15	8.6 × 10 = 86	0.101	0.558
180	1.25	7.0 × 20 = 140	0.166	0.724
200	1.39	5.0 × 20 = 100	0.118	0.842
230	1.60	2.8 × 40 = 112	0.132	0.974
270	1.87	0.5 × 40 = 20	0.024	0.998
		$\overline{845}$		

$$\bar{t} = (4/3)t_o = (4/3)(108) = 144 \text{ s}$$

The data from the above tables are plotted on Figure 11, F(t) vs. $t/\bar{t}$. Clearly these results show the significant deviation of the actual F(t) compared to that for a Newtonian fluid. The results look more like the F(t) diagrams for a power law fluid with n = 0.4, shown in Figure 7. Since soy flour doughs have n ≈ 0.2, more deviation from the F(t) for a Newtonian fluid should be expected. An inflection in the F(t) curve is seen in Figure 7 at higher values of $G_z/\cos \theta$. Because a relatively deep flighted screw was used in this example, the reduced pressure flow term should be large and may account for some of the deviation noted.

Another reason for some difference between the F(t) curves found and those previously reported could be in the value of $\bar{t}$ used in the calculation. $\bar{t}$ was estimated on the basis of a Newtonian fluid and the real relationship is given in Figure 8. Because $Q/2Q_d$ was not known, this relationship was not used. It is suspected, however, that $Q/2Q_d$ is less than 0.5 so $\bar{t}$ was probably underestimated causing a shift in the curves shown.

III. STRAIN

A. Mixing

Mixing is important within the food extruder to assure uniformity of the extrudate. To assure complete mixing, it is common extrusion practice to preblend dry and liquid ingredients before they enter the extrusion screw. In addition, mixing of the components in the feed also occurs within the screw itself. Such mixing is crucial to achieving uniformity of the dough before it leaves the die, when liquid materials, water, or steam are added in the extrusion barrel.

Mixing occurs within an extruder due to the existence of laminar shear flow. The thoroughness of mixing can be measured by two parameters: the scale of segregation and the intensity of segregation of a minor component within some major component. The scale of segregation pertains to the actual size of the minor component after mixing, while the intensity of segregation is a measure of the difference in the concentration of the minor component as compared to the desired concentration. In extrusion processes, the scale of segregation is the most useful parameter used to characterize the thoroughness of mixing, because the residence times are relatively short and diffusion is very slow in the viscous food doughs.

The reduction of the scale of segregation or the increase in mixing is related to the total strain that a material has received. Strain is defined as

$$\gamma = dx/dy \tag{5.18}$$

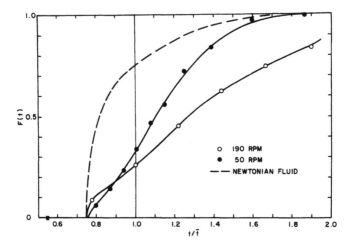

FIGURE 11.   F(t) for soy dough.

or the relative displacement of a fluid element under deformation. Shear strain can be defined as

$$\gamma = \dot{\gamma}t \qquad (5.19)$$

where $\dot{\gamma}$ is the shear rate and t is the time under which deformation occurs. Interestingly, increasing flow rate does not necessarily increase γ but does increase $\dot{\gamma}$. Strain is the product of $\dot{\gamma}$ and the residence time in the system.

Strain varies across the cross section of the channel because both the shear rate and residence time vary with initial position in the channel as previously discussed. Strain is related to the thoroughness of mixing, because a system which has received more strain has a smaller striation thickness in the mix. The combination of the strain and residence time distribution gives a unique parameter, the weighted average total strain (WATS) which can be related to mixing.[6]

B. WATS

To define WATS or $\bar{\gamma}$, it is first necessary to describe the flow field within the channel. Remember that there is a circulation within the channel with a particle having corresponding positions in the upper and lower sections of the channel, y and y_c, given by Equation 5.3. In the upper path,

$$\dot{\gamma}(y) = [\dot{\gamma}_x^2 (y) + \dot{\gamma}_z^2 (y)]^{\frac{1}{2}} \qquad (5.20)$$

where $\dot{\gamma}(y)$, $\dot{\gamma}_x(y)$, and $\dot{\gamma}_z(y)$ are the total shear rate, shear rate in the x direction and shear rate in the z direction evaluated at y, respectively. Similarly, for the path y_c corresponding to y

$$\dot{\gamma}(y_c) = [\dot{\gamma}_x^2 (y_c) + \dot{\gamma}_z^2 (y_c)]^{\frac{1}{2}} \qquad (5.21)$$

To find the strain experienced by a particle of fluid, it is necessary to multiply the shear rate by the fraction of time spent in that region or

$$\gamma(y) = \dot{\gamma}(y)t(y)t_f + \dot{\gamma}(y_c)t(y)(1 - t_f) \qquad (5.22)$$

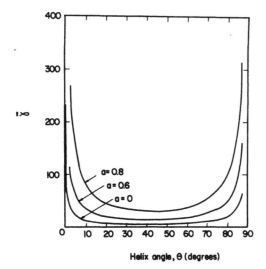

FIGURE 12. The weighted average strain in an extruder for a Newtonian liquid. (From Pinto, G. and Tadmor, Z., *Polym. Eng. Sci.*, 10(25), 279, 1970. With permission. © Society of Plastics Engineers, Inc.)

The WATS can be calculated by

$$\bar{\gamma} = \int_0^\infty \gamma(y)E(t)\,dt \qquad (5.23)$$

In the integration it is necessary to understand that the residence time varies with each particle path, or for each t there is a corresponding path y and y_c.

The analytical solution of Equation 5.23 for isothermal Newtonian flow has been developed by Pinto and Tadmor.[6] They used Equations 5.1 and 5.2 to describe the velocity distributions within the channel. Clearly v_z is a function of the relative magnitudes of the pressure and drag flow, a, and both velocities are related to the helix angle, θ. The relationship of these variables is given in Figure 12. From the figure, it can be seen that $\bar{\gamma}$ is little influenced by θ over the range of 20 to 75°. Some food extruders have helix angles between 10 and 20° and at the lower θ, $\bar{\gamma}$ would be greater. In other words, extruders having screws with lower helix angles should provide better internal mixing.

The WATS, $\bar{\gamma}$, can be normalized against a nominal strain which assumes the shear rate in the extruder is approximately V_z/H which is exerted for the average residence time, $\bar{t}$. Therefore

$$\gamma_N = \frac{\pi DN \cos\theta\,\bar{t}}{H} \qquad (5.24)$$

and the normalized strain is defined as

$$\bar{\gamma}^* = \bar{\gamma}/\bar{\gamma}_N \qquad (5.25)$$

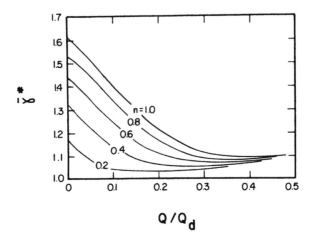

$$\frac{Q}{Q_d}$$

FIGURE 13.   Normalized strain as a function of $Q/2Q_d$ and n. (From Bigg, D. and Middleman, S., *Ind. Eng. Chem. Fundam.*, 13(1), 66, 1974. With permission. Copyright © 1974 American Chemical Society.)

Bigg and Middleman[1] have calculated $\bar{\gamma}^*$ for power law fluids as a function of both n and $Q/2Q_d$. Their results are given in Figure 13. For a fixed $Q/2Q_d$, $\bar{\gamma}^*$ is smaller with non-Newtonian fluids but the difference is greatest at very low flow rates.

The determination of $\bar{\gamma}^*$ can be done for any extrusion condition using Figure 13. First, $Q_z/\cos\theta$ given by Equation 5.17 is calculated. With Figure 14, Chapter 4, $Q/2Q_d$ is estimated. Knowing n and $Q/2Q_d$, $\bar{\gamma}^*$ can be read directly from Figure 13. No work is reported where $\bar{\gamma}^*$ is correlated with the food extrusion processes or products, but such an analysis could prove fruitful.

REFERENCES

1. **Bigg, D. and Middleman, S.**, Mixing in a screw extruder. A model for residence time distribution and strain, *Ind. Eng. Chem. Fundam.*, 13(1), 66, 1974.
2. **Bruin, S., van Zuilichem, D. J., and Stolp, W.**, Fundamental and engineering aspects of extrusion of biopolymers in a single screw extruder, *J. Food Proc. Eng.*, 2(1), 1, 1978.
3. **Griffith, R. M.**, Fully developed flow in screw extruders, *Ind. Eng. Chem. Fundam.*, 1(3), 180, 1962.
4. **Levenspiel, O.**, *Chemical Reaction Engineering*, 2nd ed., John Wiley & Sons, New York, 1972.
5. **McKelvey, J. M.**, *Polymer Processing*, John Wiley & Sons, New York, 1962, 299.
6. **Pinto, G. and Tadmor, Z.**, Mixing and residence time distributions in melt screw extruders, *Polym. Eng. Sci.*, 10(5), 279, 1970.
7. **Tsao, T. F.**, Available Lysine Retention During Extrusion Processing, Ph.D. thesis, Colorado State University, Fort Collins, 1976.
8. **van Zuilichem, D. J., Swart, J. G., and Brusman, G.**, Residence time-distributions in an extruder, *Lebensm. Wiss. Technol.*, 6(5), 184, 1973.

Chapter 6

EXTRUSION MEASUREMENTS AND EXPERIMENTATION

I. OPERATING VARIABLES

There are two types of operating variables — independent and dependent. Independent variables are those which can be controlled by the extrusion systems operator and in no way are they dependent upon other factors present within the system. In contrast, dependent variables are those which achieve a certain value which is dependent upon the magnitude or level of the independent variables. Mathematically, independent variables are functions of no other independent variable, while dependent variables are always functions of one or more independent variables.

There are relatively large numbers of variables which are affected by and affect extrusion systems. The development of extrusion models for flow, power, residence time, strain, etc. in Chapters 4 and 5 clearly point to the complex nonlinear interrelationships between extruder geometry and these process variables. Lists of independent and dependent variables should help the reader make a clearer distinction between them. Understanding the character of the variables involved can assist greatly in the operation of or experimentation on extrusion systems.

II. INDEPENDENT VARIABLES

The following is a list of independent variables which affect food extrusion.

A. Feed Ingredients

The ultimate character of the food extrudate is dependent upon the composition of the feed ingredients. It is, therefore, critical that the feed ingredients be carefully specified and controlled. In many cases this is easier said than done, because food materials are biological and can have considerable natural variability.

In characterizing feed ingredients, the supplier of all components (brand name, manufacturer, lot number) needs to be specified. Quality control should set up laboratory tests which are run on all ingredients to assure their consistency and quality. Such tests include moisture, particle size, proximate analysis, functional properties (solubility, viscosity, purity, etc.), color, and flavor.

The need to characterize extrusion ingredients cannot be overemphasized. The presence of small quantities of some materials can affect the complex cooking reactions occurring in the extruder. For example, small changes in the level of reducing sugar can affect lysine inactivation by browning, changing pH, or salt concentration altering protein denaturization.

Once the character of the individual ingredients in a food formulation is established, it is necessary that they be thoroughly mixed in the proper proportion with uniform holding time. Although the proportion and time are theoretically independent variables, the character of the materials handling equipment, scales, etc. will cause a certain amount of variability in the results.

B. Moisture

Moisture is listed as a separate variable in addition to feed ingredients because it is often controlled separately in the extruder. The amount and location of moisture additions are independent variables. Moisture can be added directly to the feed, injected into the barrel, or added in the form of steam to the preconditioner or barrel. In all these cases, the temperature of the feed material will be influenced by the type and amounts of moisture addition.

C. Screw and Barrel Design

The geometry of the extrusion screw and barrel can be changed by substituting one component for another. In some food extruders, the screws and barrels are segmented or built up from a number of pieces attached to a shaft or one another, which gives a significant amount of variability in the geometrical configuration of the extruder.

D. Die Design

The size, shape, number, and location of the dies or die inserts are all independent variables.

E. Screw Speed

Many extruders have a variable speed drive which enables the operator to change the screw speed easily and independently.

F. Jacket Temperatures

The barrel is often surrounded by hollow jackets or heaters which can be controlled to achieve a specific outer barrel temperature. In certain cases, the screw is hollow and by circulating water or steam, it can be either cooled or heated.

G. Preconditioning

Feed ingredients can be wetted and/or heated with steam in a preconditioning chamber, a continuous mixer at the entrance of the extruder. Preconditioning is best characterized by the changes in ingredient temperature and moisture that occur.

H. Feed Rate

Feed rate may or may not be an independent variable depending upon the way the extruder is operated. It is common practice to keep the feed hopper full for choke feeding so the feed rate is the quantity which the extruder will accept. Under these conditions, the feed rate is actually a dependent variable and a function of the nature of the feed ingredients, screw, barrel, and feed jacket temperatures.

Alternatively, extruders can be starve fed where the feed section is not completely full and the feed rate is less than would occur if the hopper or feeder were kept completely full. Under these conditions, the feed rate can be varied by the operator to different levels.

III. DEPENDENT VARIABLES

A number of important dependent variables for food extrusion are listed below.

A. Viscosity, η

As discussed in Chapter 3, the rheological properties of a food dough are quite complex and non-Newtonian. The viscosity is effected by composition, moisture, cooking, shear rate, and temperature.

B. Shear Rate, $\dot{\gamma}$

The shear rate occurring within the screw channel, clearance, and the die are all dependent variables. The $\dot{\gamma}$ in the screw depends upon the screw speed, N, and the geometry of the screw (Equation 4.41), in the clearance δ, (Equation 4.85), while in the die it depends upon the size and shape of the die, the flow rate, and the rheological properties of the food dough (Equations 4.78 and 4.79).

C. Flow Rate, Q
Equation 4.22 describes the volumetric flow rate as a function of extruder geometry, N, ΔP and the dough viscosity.

D. Pressure, P
The pressure at the discharge of the extruder attains a value which balances the output of the extruder with the flow through the die. The pressure drop across a die is most easily estimated by Equation 4.27, requiring knowledge of the geometrical shape of the die and the dough rheology.

E. Power, E
Equation 4.66 describes the functional relationship between the power dissipated in the screw and its geometry, N, ΔP, and the dough rheology. Power is the single most sensitive dependent variable characterizing the operation of a food extruder.

F. Residence Time, $\bar{t}$, and/or RTD
Equation 5.12 describes the average residence time as the ratio of the void volume of the screw, which comes directly from its geometry, and the volumetric flow rate, Q. The RTD can be measured using an instantaneous injection of dye, as described in Chapter 5, to obtain the E(t) and F(t) diagrams.

G. WATS, $\bar{\gamma}$
The WATS can be estimated with Figure 13 in Chapter 5 when the volumetric flow rate and n are known.

H. Temperature, T
Food dough temperature is a function of the mechanical energy input and heat transfer as described by the energy balance in Equation 4.70.

I. Product Characteristics
In the final analysis, the most important dependent variables are the product characteristics. Typical product characteristics of concern are density, texture, strength, gelatinization, fiber formation, water uptake, flavor, color, appearance, etc. No specific functional relationships have been found between product characteristics and process variables.

Once the variables of extrusion have been defined, it is necessary to find and use reliable instruments or methodologies to measure and/or characterize them.

IV. VARIABLE MEASUREMENT

Some extrusion variables require standard laboratory analysis or measurements and descriptions of sizes and shapes of machine components. No specific discussion will be given on these but rather the focus of the following material will be on process and product variables.

With certain extrusion variables, measurement and control are synonymous. Proper instrumentation is necessary to achieve uniform operations. To receive maximum benefit from experimentation with the extruder, with extension of the results to other situations, controlling, measuring, and recording of all pertinent data are essential.

A. Water Rate

The most common type of manually controlled water feeders used on extruders are those which are designed to deliver a constant volume of water. Since the density of water is nearly constant over the temperature range of interest, volumetric metering of water is satisfactory.

Several types of water metering devices used are listed below.

Water wheels — Delivery rate is varied by controlling the speed of rotation of the water wheel or the level of the water in the reservoir which fills the cups on the rotating wheel. Delivery rate is proportional to speed of rotation and water level.

Volumetric pumps — Delivery rate is related to the stroke rate of a piston pump or the length of discharge stroke varied by adjusting its attachment location on the crank shaft of the pump. Such devices are useful to the delivery of very small quantities of liquid against high back pressure.

Constant head (pressure) variable orifice meters — Water is maintained at a constant pressure and a needle valve or other variable orifice is adjusted to control the delivery rate.

Variable head (pressure) constant discharge orifice meters — The water level in a column above a fixed orifice is varied to alter the delivery rate of the orifice.

A rotameter is frequently used with the constant head variable orifice meter to indicate flow rate. A rotameter is shown in Figure 1 and is known as a variable area meter since the bob or indicating float moves in a tapered tube. As the flow increases, the bob rises, increasing the annular area between it and the tube, making the pressure differential across the bob constant.

Each of the water meters described above are relatively inexpensive, but the delivered flow rate is only indirectly sensed making them inferential meters. The flow rate in all of them is indirectly related to a speed, pressure, or an orifice opening. Should something occur to the primary elements of the devices, flow will vary and the operator has to resort to a feed check to determine if the meter calibration has shifted.

A second type of water flow meter is the feed back type diagrammed in Figure 2. Here, a primary sensing element produces a signal proportional to the actual flow rate or process signal (P.S.). This signal is fed to a controller which compares the actual feed rate to a desired rate (set point, S.P.) to give an error or deviation. Using the error signal, the controller computes a control signal (C.S.) which positions a control element, in this case a valve, so the desired flow is maintained.

Several types of primary sensing elements are used for flow measurement. The most common are described below.

Orifice plates — The square root of the pressure drop across a fixed orifice in the water line is proportonal to flow. Orifice meters are relatively inexpensive but the non-linearity of the signal and high pressure drops are disadvantages.

Venturis — Again the flow is proportional to the square root of the pressure drop, but the pressure drop is much smaller than found with orifices.

Turbine meter — A small turbine in the line revolves at a speed directly proportional to flow.

Nutating meter — A wobble disc piston fits into a chamber where the liquid flow causes the disc to nutate at a frequency proportional to flow.

In the fluid-flow feedback measurement and control systems, variable valves are the most common control element. They are reasonably inexpensive, reliable, and can give nearly linear responses over considerable range of variation.

B. Steam Rate

Much of the discussion on measurement and control of steam is similar to that al-

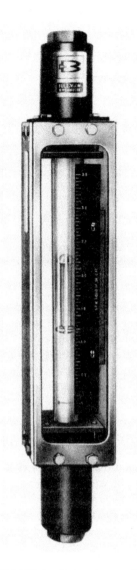

FIGURE 1. Rotameter used to measure water or steam rates to extruders. (Courtesy of Brooks Instrument Division, Emerson Electric Co., Hatfield, Pa. With permission.)

ready given previously for water rate. The feed back control systems and sensing elements are the same general types used for water. The simplest measurement and manual control system for steam involves a control valve and rotameter indicator.

C. Speed

The most common method of measuring screw speed is with a small DC tachgenerator where the voltage output is proportional to speed. Other common transducers are magnetic sensors located near the teeth of a gear on the drive shaft. The frequency of pulses from such transducers is directly proportional to screw rotational speed.

D. Feed Rate

Different types of dry ingredient meters are classified into two categories: volumetric

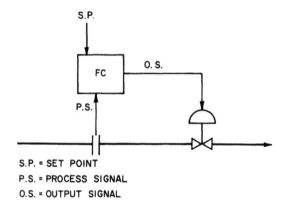

S.P. = SET POINT
P.S. = PROCESS SIGNAL
O.S. = OUTPUT SIGNAL

FIGURE 2.   Feed back flow control loop showing sensing orifice, flow controller with set point and control signal to valve.

and gravimetric. The volumetric feeders are designed to provide a constant volumetric flow rate of dry ingredients over time. For products which have a uniform density, volumetric feeders are the simplest and, therefore, the least expensive. Gravimetric feeders are designed to deliver a uniform mass rate of dry ingredients which is independent of ingredient density. Such devices usually incorporate some sort of continuous weighing belt with a load cell or weighing beam.

Volumetric feeders feature:

1. Variable speed screw conveyors where the volume of dry ingredients delivered is proportional to the speed of the screw
2. Variable speed belts having a feed hopper with a fixed opening feeding the belt; feed rate is proportional to the varying belt speed
3. Constant speed belts which have a variable dry ingredient feed opening to continuously adjust the cross-sectional area of the material falling on the belt; feed rate is proportional to the size of the adjustable opening
4. Vibratory feeders which are attached to dry ingredient hoppers and control the volumetric rate of feed delivery by the frequency or amplitude of vibration of the feeding pan as shown in Figure 3; in some cases, the vibration from the feeder may minimize bridging of ingredients in the feed hopper

Gravimetric feeders include:

1. A constant speed weight belt where the quantity of dry ingredients on the belt is varied to achieve a constant weight and, therefore, constant feed rate
2. A variable speed weight belt where the speed of the belt is changed to give a constant delivery of feed materials in proportion to the weight of ingredients on the belt at any given instant; such a gravimetric feeder is shown in Figure 4

Most dry ingredient feeders used with extruders are of the inferential type which means the actual feed rate can only be inferred from a setting on the feeder. Many times, changes in feed rate can occur because feed materials bridge in hoppers causing flow fluctuations to occur on the moving belts or in feed screws. The gravimetric feeds are, therefore, the most accurate measuring devices since a reading of the weight of material on the delivery belt is available. These devices can be attached to feed back

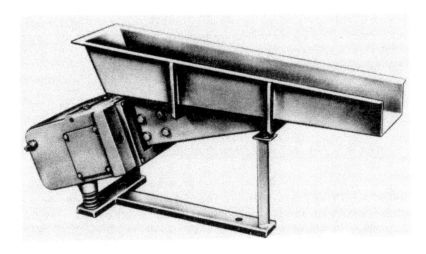

FIGURE 3. Vibrating pan metering feeder for dry ingredients. (Courtesy of FMC Corp., Homer City, Pa. With permission.)

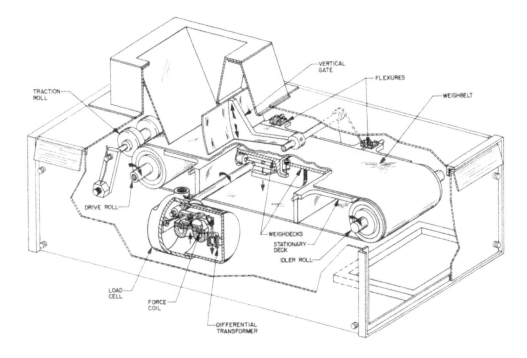

FIGURE 4. Gravimetric feeder with weight belt and variable belt speed. (Courtesy of Wallace & Tiernan, Belleville, N.J. With permission.)

controllers with the weight of material on the belt being multiplied times the belt speed to give a feed rate. When the feed rate differs from the set point, the belt speed or quantity of material on the belt can be changed to maintain a constant feed rate.

E. Flow

Flow rate at the extruder discharge is most commonly measured by taking a timed sample and weighing it. Such a system has its limitations because it is not continuous and the length of time over which the sample is taken is proportional to the accuracy of the measurement. During the collection process, moisture is usually lost and a correction of the flow is required. By taking a moisture measurement on the sample which is collected, the corrected flow can be calculated by

$$\dot{m}_1 = \dot{m}_2 \left(\frac{1 - M_2}{1 - M_1} \right) \tag{6.1}$$

where $\dot{m}$ = mass flow rate, M = moisture fraction, wet basis, subscripts 1 = in extruder, and 2 = sampling point.

It is theoretically possible to obtain continuous mass flow measurements with a continuous weigh belt or nuclear sensor. In practice, this is rarely done because of the expense of the installation and the sticky nature of the freshly extruded product.

To convert mass flow rate to volumetric flow, the correct dough density is required in the equation

$$Q = \dot{m}/\rho \tag{6.2}$$

where Q = volumetric flow rate, $\dot{m}$ = mass flow rate, and ρ = density. Density in this equation should be the density of the food dough at extrusion conditions.

F. Pressure

Pressures are not commonly measured in food extruders although they are very important. Measurements just behind the die are most useful in determining steady flow conditions and the relative magnitude of the pressure to drag flow using the flow Equation 4.22. In performing experimental work on food extruders, pressure measurements become essential. Here, it is often desirable to at least measure the pressure at the beginning of the metering section of the screw in addition to the pressure at the die head. Standardized mounting holes have been developed to receive both pressure and temperature probes and are shown in Figure 5.

The most common pressure transducer is the grease-filled, Bourdon-tube type. This is mounted through the side of the barrel and a standard grease gun, with food grade grease, is used to fill the gauge and connecting tube. Should dough plug the connecting tube, the grease gun can be used to insert additional grease and clear the passage. Bourdon-type gauges serve as good pressure indicators but are not easily adapted to electronic data collection.

The newer pressure transducers are of the diaphragm type. An example of a diaphragm type which provides a variable air signal proportional to the process pressure is shown in Figure 6. Here, a supply of air entering the transmitter passes through a regulating valve at the transducer tip. The position of the valve is controlled by the movement of a force rod so the system seeks an equilibrium. Such devices are called force balance transducers.

Other types of diaphragm pressure transducers are liquid filled. Here, the force on the diaphragm is transmitted by the liquid to a pressure transducer, which is normally instrumented with a strain gauge. A potentiometer is used to measure the change in resistance of the strain gauge, which is one leg of a bridge circuit. Some liquid-filled, pressure transducers are filled with mercury, obviously unsuitable for food extrusion. Silicone-fluid-filled gauges are available as alternates.

Diaphragm pressure transducers containing piezo-electric quartz crystal sensors are also available. They have the advantage of small size and no liquid filling.

Some pressure transducers come with a thermocouple fixed to the sensing dia-

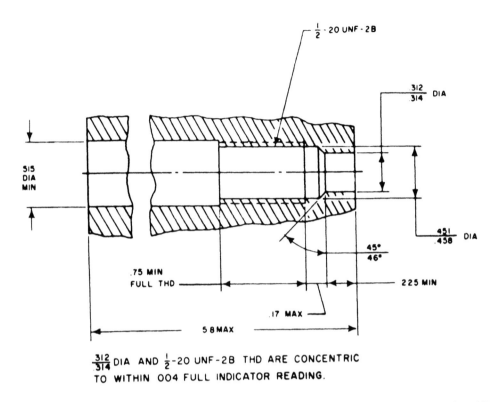

$\frac{312}{314}$ DIA AND $\frac{1}{2}$-20 UNF-2B THD ARE CONCENTRIC TO WITHIN OO4 FULL INDICATOR READING.

FIGURE 5. Standardized mounting hole for pressure thermocouples and transducers in extruders. All dimensions are in inches. (Courtesy of Dynisco, Westwood, Mass. With permission.)

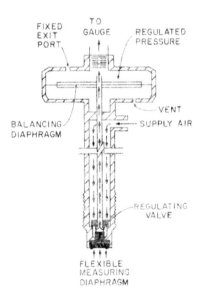

FIGURE 6. Diaphragm-type pressure transducer with variable air pressure output. (Courtesy of Rosemount Engineering Co., Bulletin 7672, Minneapolis, Minn. With permission.)

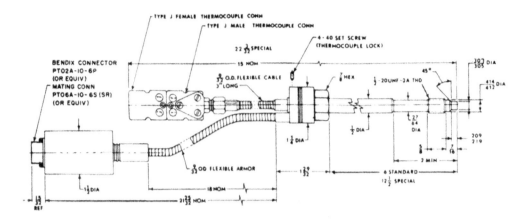

FIGURE 7.   Liquid-filled diaphragm pressure transducer combined with thermocouple. All dimensions are in inches. (Courtesy of Dynisco, Westwood, Mass. With permission.)

phragm. Such combinations allow pressure and temperature readings at the same point and save having multiple holes through the barrel. Their disadvantage is their greater expense and the fact that if one portion of the transducer fails, the entire device must be repaired. An example of such a combination thermocouple/pressure probe is shown in Figure 7.

Food extruders normally operate at pressures less than 14 MPa (2000 psi) but it is not uncommon that pressures of 65 MPa can be measured at the die of a dry extruder at conditions which reduce flow. Most pressure transducers will withstand pressures up to ten times their rated range. Using pressure transducers with too high a range will result in a loss of sensitivity in the measurements.

G. Power

A watt meter for AC drives or a combination ammeter-voltmeter for DC drives produces a continuous indication of motor load. On AC drives with no watt meter, a simple amperage measurement is preferred over no measurement. Care should be taken in the interpretation of amperage measurements because the power factor is unknown and can change, making direct power calculations difficult. Watt meters measure the total power which the drive motor draws. It is important that the no-load power — power measured when extruder is operating empty — be subtracted from the total power to get a better indication of power input to the product.

Instead of measuring power, torque and speed can be used to calculate power. Torque measurements can be obtained with a commercial foil strain gauge mounted on the input shaft with a slip ring and brush assembly to transmit the signals. Such devices are relatively expensive and subject to error because of small resistances in the slip rings, and are not frequently used.

H. Residence Time

The dye injection procedures are the best available for getting an accurate estimate of residence time in the extruder. These have been previously discussed in Chapter 5.

I. Temperature

Two types of temperature are measured for extrusion processes. One is product temperature, sometimes called melt temperature, and the other is jacket, barrel, or die temperature. Thermocouples are easily the most frequently used temperature-measur-

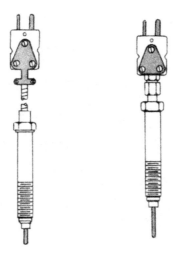

FIGURE 8. Thermocouples fitting standardized mounting hole. (Courtesy of Omega Engineering, Inc., Stamford, Conn. With permission.)

ing devices, with the change in voltage produced being proportional to the temperature of the bimetalic junction. Despite the simplicity and relative low cost of measuring temperatures with thermocouples and a multipoint sensing/recording device, accurate and reliable product temperatures are difficult to achieve.

Typically, grounded thermocouples contained in a stainless steel sheath are inserted through a hole in the side of the barrel. It is possible to purchase thermocouples with varying lengths of exposed tips, as indicated in the drawing of a rigid single element thermocouple with plug shown in Figure 8. When the temperature of the dough in the barrel is being sensed, then the exposed tip must be flush with the barrel surface or it will be damaged by the flights of the screw as they pass. When the tip of a thermocouple extends into the food dough, in areas behind the die or at the end of the screw, the tip is subjected to bending forces due to the flow of the dough. Sufficient strength in the probe is necessary or the tip can be bent or damaged.

The major problem with temperature measurement is that the probe has a relatively small surface area exposed to the food at the tip, compared to the surface area of the shaft which passes through the barrel. Heat is readily conducted in the shaft either to or from the tip so that the reading of the thermocouple usually approaches the barrel temperature. The work of Rohsenow and Choi[10] showed the temperature of the tip of the probe differing from the product temperature as

$$\frac{T_L - T}{T_b - T} = \frac{1}{\cosh(\beta L)} \qquad (6.3)$$

where T_L = temperature at end of probe, °C, T_b = temperature of barrel, °C, T = product temperature, °C, $\beta = (2h/kR)^{1/2}$, m^{-1}, h = convective heat transfer coefficient from product to probe, W/m^2°C, L = exposed length of probe, m, k = thermal conductivity of probe, W/m°C, and R = radius of probe m. Clearly, probes having a small radius and longer exposed tip length will indicate temperatures at the tip closest to the product temperature.

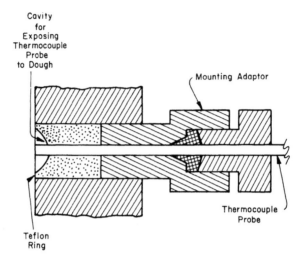

FIGURE 9.   Thermocouple used to sense temperature in a food extruder. (From Thompson, D. R. and Rosenau, J. R., *Trans ASAE*, 20(2), 397, 1977. With permission.)

Thompson and Rosenau[13] have discussed the temperature sensing problem in food extruders. For flush mounted thermocouples, the error in sensing the correct temperature can be significant. To partially overcome this problem, they suggested the use of a Teflon® plug to insulate the probe from the barrel as shown in Figure 9. These authors also analyzed the possibility of using exposed thermocouples as a way of decreasing the time response of the transducer. They concluded that the necessary mechanical strength of the exposed probe would require larger thermocouple wire and the final result would show little improvement over conventionally shielded and grounded thermocouples. Under any circumstance, the order of time response for thermocouples is approximately that of flights passing the tip, so accurate measurement of temperature profiles within the flight is difficult.

One way suggested to overcome the temperature sensing problem of the food dough is to insert thermocouples through a hollow screw and mount them flush to the root of the screw. Such a technique requires that the small thermocouple signals be brought out through a slip ring arrangement at the end of the screw which is both costly and error-prone due to contact resistence. Also, the flow at the root of the channel is slow which reduces h, the convective heat transfer coefficient, and the response time of the thermocouple.

Because of the problems with measuring product temperatures down the length of the barrel, thermocouples are sometimes only inserted into the dough at the die, and barrel temperatures are measured elsewhere. Barrel temperatures can be measured readily with surface-mounted thermocouples. Many times these are attached with a bayonet fitting which is spring loaded to assure contact between the thermocouple and the barrel surface.

J. Product Characteristics

A whole range of product characteristics can be measured. Usually this is done on samples which are collected and analyzed off line. In the later chapters of the book, descriptions of certain product tests (viscosity, gelatinization, water holding capacity, fibrization, retort stability, etc.) peculiar to specific products are given and will not be repeated here. Below is a brief discussion of some commonly used determinations of product characteristics.

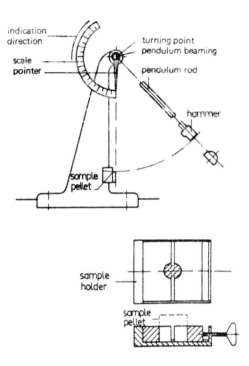

FIGURE 10. Charpy-Hammer impact resistance testing machine. (From Reinders, M. A. et al., Int. Snack Seminar, Central College of the German Confectionery Inst., Solingen, Germany, October 18 to 21, 1976. With permission.)

Density — A weighed sample of uniform particle size is placed in a large graduated cylinder and tapped to settle. The bulk density is the ratio of mass to volume. In some cases, the density of individual pieces may be of interest and is determined from the mass and volume of individual pieces.

Moisture — Moisture determinations using standard oven drying techniques for food products are used.

Strength — Shear strength is measured by placing a sample in a shear cell in conjunction with the Allo-Kramer® shear press or Instron Universal Testing Machine®. Orientation of the sample in the cell can alter the results of the test. Results are given as maximum force/unit area.

Impact — The impact resistance testing machine gives the response of extruded products to high deformation velocities. Such a device is the Charpy-Hammer testing machine, where a weight on a pendulum breaks the sample, is shown in Figure 10. The energy consumed is proportional to the loss in kinetic energy of the weight, and is reported as energy per unit cross-sectional area of sample.

Size — The size of the product can mean several things. Many times, it is the diameter of the extruded product once it is cooled. In other cases, the maximum diameter achieved after the product leaves the die is of interest. Finally, the size distribution, measured with standard sieves, may be measured.

Organoleptic properties — The properties of flavor, appearance, odor, etc. are perhaps best characterized by trained taste panels. The results of these panels can be quantitized and used as a reproducible measure of organoleptic character. Details of such procedures are beyond the scope of this book and the reader is referred to the work of Amerine et al.[2] and Stone et al.[11]

Texture — Texture is a combination of properties including fracturability, hardness, cohesiveness, adhesiveness, springiness, gumminess, and chewiness. The science of measuring texture has grown extensively with the development of instruments like the General Foods Texturometer®. Details of such testing are given in Bourne.[3]

Color — Color measurements have been most precisely done using the Hunter Tri-stimulus Color Difference Meter®. Such meters allow color to be described in terms of L (total lightness), A (+ = red, − = green), B (+ = yellow, − = blue), and ΔE (total color difference). In using such meters, it is important that a standard color be specified and the meter set to that standard before each determination. Routine testing may not require the sophistication of the color difference meter and simpler reflectance meters are are suitable.

Cook — The amount of cooking occurring in an extruder is difficult to define. Starch gelatinization measured by increased enzyme susceptability is one example of a determination to estimate the amount of cooking. Changes in product viscosity and texture can be others.

The characterization of the important variables and their measurement and control is fundamental to consistent and steady operations in production. Experimentation by engineers/food scientists is necessary to understand how changes in the independent variables affect the dependent variables. In the following section, experimental procedures for food extruders will be discussed along with a brief description of correlating procedures for experimental data including response surface analysis.

V. EXPERIMENTATION

An understanding of the effect of independent process variables in the extrusion operation and product characteristics is very important in food extrusion. The use of designed experiments to systematically gather processing data has been successfully applied[1,6-8,12,15] to food extrusion. The resulting least squares regression models are a useful way of correlating data and describing the effects of the changes on independent variables. Such techniques can describe the operating characteristics of a single machine but extreme caution should be used in the extrapolation of these results to other extruders or circumstances. The relationship between the operation of different extruders can best be determined through the application of the mechanistic models for flow, energy, residence time, WATS, etc., as described in Chapters 4 and 5. Models described in these chapters clearly define the relationships between geometrical and operating parameters. No mechanistic models describing the effect of extrusion conditions on product characteristics have yet been developed.

VI. DESIGNED EXPERIMENTS

An experimental design is a formalized experimental plan. In this plan, the independent variables and their levels are preselected. The experimentation is performed so that more than one variable is varied at a time in random order. Analysis of the resulting information not only provides the first order (single variable) effects but also gives the effects of the interaction of variables. Henika[5] has given simple examples of the effective use of designed experimentation, with statistical analysis of the results applied to food processing systems.

In selecting factors and their levels for designed experiments, prior knowledge of the important variables and their acceptable levels must be known. To reduce the number of experimental trials, the fewer the number of variables the better. The levels of the variables must also be selected so that operation of the process is feasible.

TEST	X_1	X_2	X_3
1	-1	-1	0
2	1	-1	0
3	-1	1	0
4	1	1	0
5	-1	0	-1
6	1	0	-1
7	-1	0	1
8	1	0	1
9	0	-1	-1
10	0	1	-1
11	0	-1	1
12	0	1	1
13	0	0	0
14	0	0	0
15	0	0	0

FIGURE 11. Fractional factorial experimental design for three variables at three levels. (Courtesy of Henika, R. G., *Cereal Sci. Today*, 17, 309, 1972. With permission. © 1972 American Association of Cereal Chemists.)

A variety of designs have been used successfully. A full factorial design consists of a set of experiments where all combinations of variables at their different levels are run. In the case of experimentation on extruders, which have a large number of variables, such experimental approaches lead to many individual runs.

One way to reduce the number of experiments is to use fractional factorial designs. For example, a 3 factor, 3 level, full factorial experiment would require 3^3 or 27 separate trial runs. In a fractional factorial design shown in Figure 11, only 15 tests are required. When more variables are involved, the reduction in experimental runs is larger. What is sacrificed in such fractional designs is the inability of the statistical least squares model to determine the third or higher order interactions of independent variables. In many cases, these are not significant and in the statistical evaluation, their sum of squares is lumped in the error sum of squares term used in the tests for significance of the remaining terms.

A number of alternative statistical designs, including fractional factorial designs for large numbers of variables, are given in Davies.[4] The reader is referred to this book for a detailed treatment of the subject. Proper selection of the experimental design will require only the minimum amount of experimentation to achieve the desired level of understanding or accuracy. Through proper selection of the design, sufficient data can be gathered so that predetermined higher orders of interaction can be estimated while others are not possible because they will be confounded with the desired interactions. Such selections require prior knowledge of the system and its response if they are to be made correctly and used most beneficially.

VII. ANALYSIS OF RESULTS

Representative samples of product are collected and all dependent variables are read and recorded after the extruder has been stabilized at each individual set of experimental conditions. It is very important that the extruder be at an equilibrium condition before samples are taken and data are recorded. For large extruders, this will probably require a minimum of 15 min between samples when rather small processing changes are made. Uniform and representative operation can usually be determined by monitoring the stability of the operating variables, particularly power input, and examining the character of the extrudate.

A considerable amount of time may be required to equilibrate the operation of an extruder after initial start up. An exact time requirement cannot be given because of the complex nature of the process. The operator should realize that it takes time for the large mass of the extruder to achieve thermal equilibrium, a necessary prerequisite for representative production. It is not uncommon for the extruder to slowly move to the equilibrium condition after about 1 hr of "stable" operations.

The analysis of the extruded samples is performed after the samples have been collected. If the samples are wet, care needs to be taken that they do not lose moisture or deteriorate through microbial action. The specific types of analyses made are a function of the experiments objectives but some suggestions have been made in the section on dependent variables.

Once all samples have been analyzed, the data can be key punched on data cards for statistical analysis. It is convenient to use one card for each experimental run, first listing all the independent variables followed by the measured dependent variables. This set of cards becomes the data deck for the statistical analysis programs used.

VIII. DATA ANALYSIS

Various types of statistical analysis for data from designed experiments are common. Standard computer programs for these analyses are available, which can be used with the data deck, and are available at all computing centers. Initially, it is common to run an analysis of variance (AOV) for each dependent variable with all the independent variables. Such an analysis will quickly show those effects which are significantly correlated with the changes in the independent variables and their interactions.

Statistical model building is also a common analysis technique because it describes how changes in the dependent variables (y) are related to changes in the independent variables (x). The functional response can be written as

$$y = f(x_1, x_2, x_3, \ldots, x_n) \qquad (6.4)$$

where y is the response function (Davies[4]).

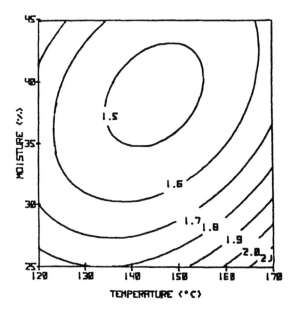

FIGURE 12. Two-dimensional response surface contour plot. Water absorption = f (moisture, temperature). (Reprinted from Aguilera, J. M. and Kosikowski, F. V., *J. Food Sci.*, 41, 647, 1976. Copyright © by the Institute of Food Technologists. With permission.)

The statistical model can take on many forms but it is convenient to use an equation which requires the fewest number of terms while providing good correlations. For the fractional factorial, three variable, three level experiment described in Figure 11, the statistical model

$$y = b_0 + b_1 x_1 + b_2 x_2 + b_3 x_3 + b_{11} x_1^2 +$$
$$b_{22} x_2^2 + b_{33} x_3^2 + b_{12} x_1 x_2 + b_{13} x_1 x_3 + b_{23} x_2 x_3$$

(6.5)

called the Taylor second order equation, is satisfactory. The term b_0 is the center point and the other b elements are the regression coeffcients that describe the changes in y with linear or first order changes in the independent variables, then quadratic or second order changes, and finally first order interactions. Various forms of regression analysis can be used to estimate the b coefficients in Equation 6.5 to minimize the residual error in the model. The F-test is normally used to determine the significance of the coefficients and only those coefficients which have $p < 0.05$ should be retained.

The model given in Equation 6.5 is called linear because it has a linear combination of terms with constant coefficients. Nonlinear regression models can also be fit statistically but estimation of coefficients usually requires more computer time and in some cases, prior information about the value of the coefficients is necessary to assure convergence. Various models and their applicability are also given in Davies.[4]

IX. RESPONSE SURFACE (RS) ANALYSIS

Once a correlation model has been developed, it is possible to develop response surfaces to help visualize the relationships between the dependent and independent variables. RSs are sometimes called contour plots which can be made either in two or three dimensions. An example of a two-dimensional RS is given in Figure 12 showing

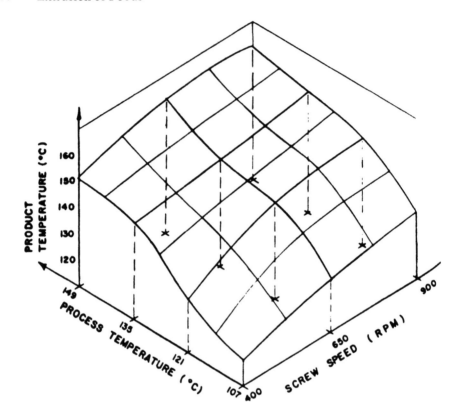

FIGURE 13.   Three-dimensional response surface shown in two dimensions. Product temperature = f (process temperature, speed). (Reprinted from Tranto, M. V. et al., *J. Food Sci.*, 40, 1264, 1975. Copyright © by the Institute of Food Technologists. With permission.)

how the variation in two independent variables effects a third dependent variable. If the model contains more than two independent variables, such a plot can still be made while all other variables are fixed at preset levels. A series of RSs can be made by changing a third independent variable, and these can be overlaid to give a three-dimensional representation of the dependence of the variables.

Actual three-dimensional representations of RS can be made in two dimensions as shown in Figure 13. The graphical representations shown were all generated directly with the computer once the model was developed, using standard graphical software available at most computing centers.

Using RS, it is possible to develop a clear mental picture of the relationships of the important process parameters, the process itself, and product characteristics. Many times, the nature of the response curves are such that it is possible to select optimal conditions for the process operation and achieve the best product characteristics using visual approaches. Nonlinear mathematical optimization techniques can also be applied, and optimal conditions can be determined considering all the process variables and their limits of variation or appropriate operating ranges.

X. LIMITATIONS OF THE RS APPROACH

The applications and benefits of the use of designed experiments to develop regression models for process understanding and optimization have been described. The models developed can correlate measured variation within the range of variables tested but should not be used to extrapolate or extend the results to regions where experimentation was not performed. Consequently, the use of such models is only applicable to the equipment used in the testing and cannot generally be transferred to other equipment.

Such a limitation means that regression models have no use in estimating the size and operation of full scale extrusion processes from data taken on laboratory scale extruders. If such information is required, it is best to consider mechanistic models of the type developed and discussed in Chapters 4 and 5. The characteristics of extruded products are dependent upon shear rates in the channel and clearance, residence distribution, WATS, temperature, energy inputs (heat transfer and viscous dissipation), etc. Experiments, where these variables are controlled and varied through the operation and construction of a series of extruders of varying sizes, would be necessary to obtain sufficient information for scale-up.

The work of Tsao et al.[14] used a set of designed experiments on an extruder with screws of different design. They then used the mechanistic models for flow and energy to correlate these experimental data. Constants in these equations were adjusted so that the model form fit the data and, therefore, better adapted itself to extrapolation. These constants represented corrections to the mechanistic equations which were necessary because of unknown factors.

No mechanistic models exist which correlate changes within the extruded product as a function of extrusion parameters. No real experimental work has been performed which would give the experimental parameters best used to correlate these changes, allowing more accurate scale-up. Currently, maintaining shear rates, temperatures, ratios of energy inputs to flow, residence times, and WATS during scale-up appears to be the best first approximation. Because of the very nonlinear nature of mechanisms governing these factors within the extruder, absolute adherence to these rules is impossible, and the optimal scale-up procedures are yet unknown and are probably a function of the system and extruder.

The last limitation to the models developed from statistical experiments is their insensitivity to long term variations in the independent variables such as ingredient variations, ambient conditions, and machine wear. These types of data can be best determined by accurately characterizing raw ingredients and measuring changes occurring within the extruder and surroundings over time. Theoretically, these could be correlated to changes in product and extruder operation. Unfortunately, this is rarely done because changes due to instrumentation drift or calibration, incomplete or inaccurate records, make such correlations difficult.

REFERENCES

1. **Aguilera, J. A. and Kosikowski, F. V.**, Soybean extruded product: a response surface analysis, *J. Food Sci.*, 41, 647, 1976.
2. **Amerine, M. A., Pangborn, R. M., and Roessler, E. B.**, *Principles of Sensory Evaluation of Food*, Academic Press, New York, 1965, 350.
3. **Bourne, M. C.**, Texture profile analysis, *Food Technol. (Chicago)*, 32(7), 62, 1978.
4. **Davies, C. L.**, *Design and Analysis of Industrial Experiments*, Oliver and Boyde, Ltd., Edinburgh, 1963, 353.
5. **Henika, R. G.**, Simple and effective system for use with response surface methodology, *Cereal Sci. Today*, 17, 309, 1972.

6. **Lawton, B. T., Henderson, G. A., and Derlatka, E. J.,** The effects of extruder variables on the gelatinization of corn starch, *Can. J. Chem. Eng.*, 50(4), 168, 1972.

7. **Lorenz, K., Welsh, J., Norman, R., Beetner, G., and Frey, A.,** Extrusion processing of triticale, *J. Food Sci.*, 39, 572, 1974.

8. **Maurice, T. J. and Stanley, D. W.,** Texture-structure relationships in texturized soy protein. IV. Influence of process variables on extrusion texturization, *Can. Inst. Food Sci. Technol. J.*, 11(1), 1, 1978.

9. **Reinders, M. A., Crum, M. G., and van Zuilichem, D. J.,** Developments and Applications of Starch Based Ingredients in the Manufacture of Extruded Foods, presented at Int. Snack Seminar, Central College of the German Confectionery Inst., Solingen, Germany, October 18 to 21, 1976.

10. **Rohsenow, W. M. and Choi, H.,** *Heat, Mass and Momentum Transfer,* Prentice-Hall, Englewood Cliffs, N.J., 1961, 94.

11. **Stone, H., Sidel, J., Oliver, S., Woolsey, A., and Singleton, R. C.,** Sensory evaluation by quantitative description analysis, *Food Technol. (Chicago)*, 28, 24, 1974.

12. **Taranto, M. V., Meinke, W. W., Carter, C. M., and Mattil, K. F.,** Parameters affecting production and character of extrusion of texturized defatted glandless cottonseed meal, *J. Food Sci.*, 40, 1264, 1975.

13. **Thompson, D. R. and Rosenau, J. R.,** Data acquisition from an extruder for food research, *Trans. ASAE*, 20(2), 397, 1977.

14. **Tsao, T. F., Harper, J. M., and Repholz, K. M.,** The effects of screw geometry on operational characteristics, *AIChE Symp. Ser.*, 74(172), 142, 1978.

15. **Walsh, D. E., Ebeling, K. A., and Dick, J. W.,** A linear programming approach to spaghetti processing, *Cereal Sci. Today*, 16(11), 385, 1971.

Chapter 7

EXTRUSION EQUIPMENT

I. INTRODUCTION

Extruders can be classified in a variety of ways. Some involve differentiation based on general functional characteristics related to the types of products produced. Other classification systems are based on thermodynamic considerations. Finally, extruders have been classified on the basis of the moisture content of the feed ingredients. All these methods are summarized below.

II. FUNCTIONAL CHARACTERISTICS

The systematic classification of food extruders based on their functional characteristics and the types of food products produced was first done by Rossen and Miller.[6] A modified version of their classification is given in Table 1 with ranges and conditions being more specifically defined. A description of the categories of extruders developed follows.

A. Pasta Extruders

The deep flighted screw (generally smooth barrel) and low screw speed of the pasta extruder make it ideal for working moistened semolina flour and pressing it through a die with little or no cooking. In the pasta extruder, minimal input energy is dissipated because of the low shear rate in the product when a smooth barrel used. Additionally, similar extruders are used to handle pastry doughs and cookies. They have also been adapted for dual extrusion of filled-food items such as egg rolls and ravioli.

B. High-Pressure Forming Extruders

Many preshaped food products are made by extruding pregelatinized cereal doughs through a die to first form unpuffed pellets which are subsequently partially dried and puffed in a frier, puffing gun, or roaster. To produce high pressures, these extruders normally have grooved barrels to prevent slip at the wall and greater compression in the screw design, which results in their ability to product higher pressures at the die. Excessive temperatures of the dough in the screw can lead to unwanted puffing at the die and unsatisfactory performance. Excess heat is removed with water circulating in a hollow screw or in jackets around the barrel.

C. Low-Shear Cooking Extruders

Many soft-moist pet foods or high moisture food products are prepared on extruders which have moderate shear, high compression to enhance mixing, and grooved barrels to prevent slip at the barrel wall. Heat can be applied to the barrel or screw to heat the product since little viscous dissipation of mechanical energy occurs due to the relatively low viscosity of the materials being extruded.

D. Collet Extruders

Rapid viscous dissipation of the mechanical energy input to the short screw (L/D $\sim$ 3:1) occurs in the collet extruder because of the relatively high shear in the shallow flights of the screw operating in a grooved barrel to prevent slip at the walls. Collet extruders normally extrude relatively dry feed materials, and heat them rapidly to tem-

Table 1

TYPICAL OPERATING DATA FOR FIVE TYPES OF FOOD EXTRUDERS

	Typical operating data				
Measurement	Pasta extruder	High-pressure forming extruder	Low-shear cooking extruder	Collet extruder	High-shear cooking extruder
Feed Moisture (%)	31	25	20—35	12	20
Product moisture (%)	30	25	15—30	2	4—10
Maximum product temperature (°C)	52	80	150	200	180
D/H	3—4	4.5	7—15	9	7
p	1—2	1	1	2—4	1—3
N (rpm)	30	40	60—200	300	350—500
Shear rate (s^{-1}) in screw	5	10	20—100	140	120—180
Mechanical energy input, kW-hr/kg	0.05	0.05	0.02—0.05	0.13	0.14
Portion of mechanical energy input dissipated as heat, kW-hr/kg	0.03	0.04	0.02—0.03	0.10	0.10
Heat transfer, q, from barrel jackets, kW-hr/kg	(0.01)[a]	(0.01)	0.04	0.0	0—(0.03)
Net energy input to product, kW-hr/kg	0.02	0.03	0.06—0.07	0.10	0.10—0.07
Product types	Macaroni	RTE cereals, 2nd generation snacks	Soft-moist products, standard soup bases	Puffed snacks	Textured plant protein, dry cereals, dry pet foods

() mean heat is transferred from product to jacket.

Adapted from Rossen, J. L. and Miller, R. C., *Food Technol. (Chicago)*, 27(8), 46, 1963.

peratures exceeding 175°C so the starch is gelatinized and partially dextrinized. When this material exits the die the rapid change in pressure and normal forces from the non-Newtonian dough cause substantial expansion of the cooked piece, resulting in a concurrent loss of moisture and the formation of a crisp, expanded curl or collet. The most common ingredient extruded is degerminated corn grits.

E. High-Shear Cooking Extruders

These extruders were designed to produce a large variety of precooked, gelatinized or heat treated food products having their tactile components restructured. Initially, plastic extruders with long barrels (L/D = 15 to 20:1), high compression ratios, and the ability to heat or cool the product externally through the barrel were used. Such extruders provide a significant operating capability, being able to accept a wide range of initial product moistures and ingredients along with an ability to control the desired processing conditions such as temperature and puffing. Product applications include pet foods, RTE cereals, textured plant protein, and snacks.

High-shear cooking extruders have been classified as HTST (high temperature/short time devices). In most applications, feed ingredients are preheated with steam or hot water and then processed through the high-shear cooking extruder to further work the product and to increase its temperature rapidly. Nearly instantaneous cooling occurs once the product leaves the die, which when coupled with the short residence time in the screw, results in the HTST designation.

III. THERMODYNAMIC CHARACTERISTICS

A thermodynamic classification of single screw extruders was also developed by Rossen and Miller[6] and is discussed below.

A. Autogenous Extruders

The entire heat input to the extruder comes from the viscous dissipation of mechanical energy inputs, and little or no heat is added or removed from the barrel. Collet extruders and some high-shear cooking extruders are examples of autogenous extruders. Since temperatures are controlled by feed composition and screw configuration, autogenous extruders tend to have less flexibility and are more difficult to control.

B. Isothermal Extruders

Constant temperatures are maintained throughout the length of the barrel of an isothermal extruder. Forming extruders typically fall in this category. To maintain isothermal conditions, heat is removed through jackets surrounding the barrel. Since dough conditions remain relatively constant in isothermal extruders, they lend themselves to easier description mathematically.

C. Polytropic Extruders

In reality, all extruders are polytropic, although some operate approximately as autogenous or isothermal extruders. Cooking extruders which have jacketed barrels, where heat is alternatively added or extracted, operate in the polytropic regime.

IV. MOISTURE CHARACTERISTICS

The moisture content of the feed ingredients greatly affects the operation of extrusion equipment and the finished product characteristics. A classification of extruders made on this basis is given in Table 2. Dry extrusion is characterized by very high mechanical energy inputs, relatively few product shapes beyond highly puffed or expanded pieces, higher maintenance costs, and a range of capital costs depending upon the capacity of the system and specific extruder. Intermediate to high moisture extrusion systems tend toward the opposite end of this spectrum, except for cost, which again is a function of extruder capacity.

Operating costs of both high and low moisture extrusion systems tend to be related to the capacity of the system, with higher capacity systems having lower operating costs per unit of throughput. Operating costs are also a complex function of energy, equipment, and labor costs, which make any generalization difficult. For a given product type, capacity and known operating variables, optimal extrusion systems, and operating conditions can be determined. The effects of operating moisture on energy costs is discussed further in Chapter 8.

V. COOKING EXTRUDERS (SINGLE SCREW)

Several manufacturers supply cooking extrusion equipment which differs in mechanical features and operating characteristics. A description of the equipment along with illustrations are given so that the reader has a clearer understanding about their distinguishing features and applications. These discussions are given below by manufacturer, in alphabetical order.

A. Anderson-Ibec

Anderson-Ibec (Division of International Basic Economy Corp., Strongsville, Ohio)

Table 2

**CLASSIFICATION OF EXTRUSION EQUIPMENT ON THE BASIS OF THE
MOISTURE CONTENT OF THE FEED INGREDIENTS**

Feature	Low moisture	Intermediate moisture	High moisture
Ingredient moisture	M ≤ 20%	28% > M > 20%	M ≥ 28%
Source of input energy	Most energy input from viscous dissipation of mechanical energy input	About half of energy input comes from viscous dissipation of mechanical energy input, other half comes from steam injection or heat transfer	Majority of energy comes from steam injection and heat transfer with very little energy coming from conversion of mechanical energy to heat
Mechanical energy	0.10 kW-hr/kg	0.04 kW-hr/kg	<0.02 kW-hr/kg
Product drying	Minimal requirement; product cooling results in 6% moisture loss	Some product drying required to remove moisture in excess of 12% in finished product	Extensive product drying is required to reduce finished product moisture
Product shape	Minimal number of shapes available beyond highly expanded pieces or flakes	Many product shapes available	Maximum flexibility of product shapes and textures
Product density	Low density expanded product	Moderate density	Range of density possible
Ingredients	Low moisture cereals and/or oil seeds	Few limitations	Few limitations
Capacity	Lower range, 0.1—0.8 MT/hr	Moderate range, 0.2—3.6 MT/hr	Wide range, 0.4—10.0 MT/hr
Capital cost (extruder, boiler, and dryer/MT)	Low to high, $60,000—$350,000/MT	Moderate, $80,000—$300,000/MT	Highest capital cost, $90,000—$270,000/MT
Maintenance costs	High, $1.20/MT	Moderate, $0.50—$0.60/MT	Low to moderate, $0.40—$0.50/MT
Operating costs	Variable	Variable	Variable

From Harper, J. M., *Low-Cost Extrusion Cookers,* Wilson, D. W. and Tribelhorn, R. E., Eds., Colorado State University, Fort Collins, 1979. With permission.

began building screw presses called expellers for the continuous extraction of oil in the early 1900s. These presses first found application in the extraction of tallow from meat scraps but were expanded to oil extraction and dewatering of paper pulp. In the early 1950s, Anderson-Ibec redesigned the expeller to heat and extrude grain for animal feeds. The first machines were marketed in 1955 and were called Vertical Grain Expanders. The horizontal extruder-expander, similar in configuration to present day designs, was introduced in 1958 with modifications and improvements being made to increase the longevity and capacity of the equipment.

The Anderson-Ibec extruder has a unique design to achieve HTST cooking. A cross-sectional drawing of the Anderson-Ibec horizontal extruder is shown in Figure 1. The screw has deep discontinuous flights and a 12:1 L/D ratio. The restriction or compression in the screw is achieved through the placement of hardened breaker bolts through the barrel, which match the cuts in the flights so as not to interfere with the rotation of the screw. Mixing within the product is caused by the addition of more breaker bolts. The screw is assembled with heat-treated screw sections slipped over the shaft and fixed into place. The screw sections at the discharge end are most subject to wear, so they are made from wear-resistant materials which can be easily replaced when worn to maintain extrusion efficiency.

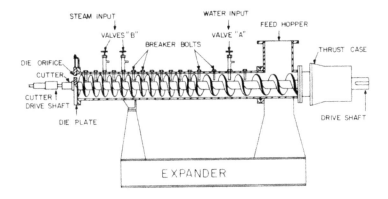

FIGURE 1. Cross-sectional drawing of 8-in. (20.3-cm) Anderson-Ibec extruder. (Courtesy of Anderson-Ibec, Strongsville, Ohio. With permission.)

Hardened stainless steel liners slip inside the barrel and are held in place with the breaker bolts through the barrel wall. These provide a replaceable, hardened-wear surface which reduces maintenance costs. The barrel on most Anderson-Ibec extruders is not jacketed. Jackets are expensive to construct, so the design reduces capital costs of the equipment. Clearance between the screws and barrel liners is about 3 mm.

The energy input into the product on an Anderson-Ibec extruder is a combination of the dissipation of mechanical energy and direct steam injection. The amount of mechanical energy dissipated is a function of the number, length, and location of the breaker bolts placed into the sides of the barrel. Obviously, more breaker bolts extending to the root of the screw will cause increased mechanical energy inputs. Some of the bolts can be hollow and fitted with a control valve to serve as steam or water injectors. High pressure steam can be injected directly into the product at the root of the screw to add significant quantities of heat and moisture.

The Anderson-Ibec extruders have no preconditioning equipment. The dry, free-flowing ingredients are easily added through the large-mouthed hopper supplied on the equipment by using a screw-type volumetric feeder. Water at 0.7 MPa is added to the product through an injection valve at a distance of about 4 (extruder) diameters from the feed hopper. The screw conveys the moistened product and works it into a dough mass. Normally, steam is injected into the product in two or three locations within the last 5 diameters of the length of the screw. Potable sparge steam having 100% quality and a pressure of 10MPa is required. Injection of the steam at the root of the screw greatly improves its incorporation into the product and reduces the chance for blow back when the steam is not condensed in the product. The quantity of water added by the condensing steam has to be considered in determining the desired moisture of the cooking operation. The drive of some Anderson-Ibec extruders is connected to the screw with a V-belt reducer. Such a system provides overload protection in case of plugging or blockage of the screw. Others have gear reduction drives.

A face cutter is attached directly to a spindle on the die plate mounted at the discharge of the barrel. The die plate and die inserts, cutter head and various cutting knives are shown in Figure 2. A variable speed drive attached to the cutting head provides control over the length of the extruded product.

Anderson-Ibec makes a range of extruders to meet a variety of production needs. Their equipment is summarized in Table 3. The rated capacity is greatly influenced by the moisture of the extruded products because specific power decreases rapidly as extrusion moisture increases.

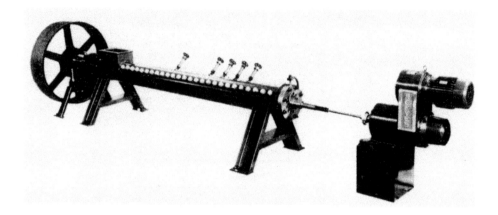

FIGURE 2.   Anderson-Ibec extruder showing die head and variable speed cutter and drive. Note water and steam injectors on barrel. (Courtesy of Anderson-Ibec, Strongsville, Ohio. With permission.)

A variety of products are being extruded on the Anderson-Ibec equipment. These include pregelatinized cereals/starches, textured plant protein, pet foods, and expanded cereals and snacks.

B. Bonnot

The Bonnot Co. (Kent, Ohio) has a long history in the extrusion field where they started by building extruders for clay products. For over 15 years, they have been actively involved in the design and development of food extruders.

The Bonnot cooking extruder has a design quite similar to plastics extruders where the screw has definite feed, transition, and metering sections. The typical Bonnot screw has a feed section which extends two flight turns into the barrel beyond the feed opening and a metering section for half the length of the screw. With this design, the transition section is quite long and has a gradual increase in root diameter to prevent binding of the product within the channel. Most of the Bonnot cooking extruders have a 20:1 L/D ratio, single flighted screw with a constant pitch of 12 to 15°, and a compression ratio of 3.5:1. Shorter cooking times are achieved with screws having shorter L/D ratios and custom designed screws are also available.

A Bonnot cooking extruder is shown in Figure 3. The barrel is made of sections which are jacketed for the addition of water or steam. Typically, cold water is used on the jacket in the feed section to aid feeding and steam of varying temperature is used on the rest of the jacketed barrel sections. To increase heat exchange area, some Bonnot screws are hollow with a quill insert and a rotating connection for steam or water attached after the screw slips through the thrust bearing.

Stainless steel is used throughout the construction of the Bonnot extruder. The screws are made from Armco® 17-4, a stainless steel having good strength, wear resistance, and machinability. To give additional wear resistance, a stellite leading edge is applied to the flight, ground and polished. Barrel liners are centrifically cast, 400 series stainless steel and heat treated to 50 to 55 Rockwell-C hardness. Very shallow 1.5-mm, longitudinal grooves in the barrel reduce slip at the walls. Typical clearances between the flight and the barrel of 0.13 mm must be maintained to achieve maximum productivity.

The amount of heat coming from the dissipation of mechanical energy is a function of the product and its viscosity. Low moisture conditions result in high mechanical energy inputs while with high moisture foods (35 to 40%), less than 20% of the energy

Table 3
DESCRIPTION OF THE AVAILABLE MODELS OF THE ANDERSON-IBEC EXTRUDER

Nominal size (in.)	Screw diameter (cm)	Length of barrel (cm)	Motor (kW)	Capacities (MT/hr)		
				Low density snacks	Medium density pet foods	Textrued plant protein
4.5	11.4	137	18.6	0.19	0.36	0.23
6.0	15.2	183	37.2	0.45	0.91	0.54
8.0	20.3	244	93.1	1.36	3.63	1.47
10.0	25.4	305	186.0	2.50	7.26	2.95

FIGURE 3. Bonnot 8-in. (20.3-cm) cooking extruders. Note relatively long L/D and jacketed barrel sections. (Courtesy of Bonnot Co., Kent, Ohio. With permission.)

Table 4

DESCRIPTION OF THE AVAILABLE MODELS OF THE
BONNOT EXTRUDER

Nominal size (in.)	Screw diameter (cm)	Screw length (20:1) (cm)	Drive motor (kW)	Nominal capacity (MT/hr)
2.25	5.72	114	14.9	0.05
4.0	10.2	204	29.8	0.27
6.0	15.2	304	112.8	0.82
8.0	20.3	406	186.0	1.36
10.0	25.4	508	298.0	2.52

addition comes from mechanical dissipation. Most Bonnot extruders come with a variable speed drive to adjust extrusion rate and energy dissipation but typically operate at <100 rpm.

Interchangeable dies can be fitted at the extruder discharge to achieve a variety of product shapes. Bonnot can supply different types of variable speed cutters to suit the product needs, including offset fly knives, rotating knives, and guillotine cutters.

The Bonnot extruders do not have any preconditioning chambers or attachments. They can supply paddle mixers with adjustable pitch blades and variable speed drives for dry and wet ingredient blending. The feed throat on the extruders can be fitted with a live bottom feeder consisting of a multiplicity of close fitting and rotating screws. The purpose of the live bottom feeder is to uniformly feed wet or otherwise sticky food materials that tend to bridge in the feed transition.

Bonnot extruders come in a large variety of sizes. A description of these extruders, capacities and power requirements, is shown in Table 4. These extruders have been used to manufacture a number of products including RTE cereals, second generation snack pellets, textured plant protein, pregelatinized starches, and soft-moist pet foods. The company maintains a test facility for customer testing of extruder applications.

C. Sprout-Waldron

The Sprout-Waldron Co. (Metal Products Div., Koppers Company, Inc., Muncy, Pa.) has been manufacturing cooking extruders since 1958. A cross-sectional diagram of their cooking extruder is shown in Figure 4.

A significant amount of the cooking of the feed ingredients is done in the pressure cooker mounted on top of the drive and extrusion screw. To maintain above atmospheric pressure in the pressure cooker, a patented, variable speed screw feeder is used with a counter-rotating seal plate which is air loaded to maintain the seal. The single shaft within the pressure cooker has radial paddles to tumble and convey the product. The shaft rotates between 10 to 24 rpm and gives a residence time of approximately 3 min. Low pressure steam and/or water can be added to the pressure cooker to give a uniformly cooked and moistened product. The precooker and all other parts in contact with the food are made from stainless steel.

The precooked product from the pressure cooker moves into the extrusion screw through a vertical transition, or it can be force fed with a feed screw in certain instances. The extrusion screw is constructed of 304 stainless steel (SS) with single flights in the feed section followed by double flights about two thirds of the way down the screw. The screws are characteristically deep flighted with a helix angle of 13° and an L/D of 8:1. The major functions of the extrusion screw are to serve as a pressure lock at the outlet of the pressure cooker, add the additional heat necessary for cooking, mix the product to assure uniformity, and force the product through the die. On cer-

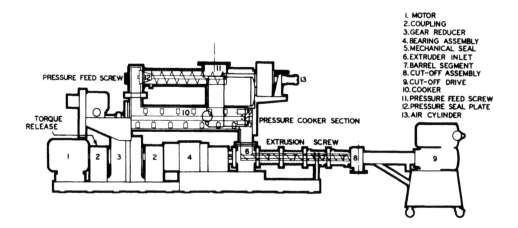

FIGURE 4.   Cross-sectional drawing of Sprout-Waldron Model 775 cooking extruder. (Courtesy of Sprout-Waldron, Muncy, Pa. With permission.)

tain screws, partial pins are placed in the screw channel near the discharge to enhance mixing and energy dissipation. Typical extrusion conditions are 33% moisture and approximately 0.7 MPa pressure.

The extrusion screw is driven at constant speed of 280 rpm with a directly coupled motor through an inline gear reducer. A torque limiting coupling is used as a safety device for the drive train and screw. The screw clamps directly to the drive shaft.

A segmented barrel has no liners and in most configurations, no jackets. The segments are made from cast 431 SS having a 38 to 42 Rockwell-C hardness. The internal surface can have either straight or spiral grooves.

A die holder assembly bolts directly into the last barrel segment, 0.5 cm from the end of the screw. Nontapered die inserts fit into the holder from the rear. A face cutter mounts to a spindle on the die holder assembly and is driven by a variable speed drive attached with a shaft and two universal joints. Multiple tool steel knives meet the face of the die at a 5° angle. The cut product falls directly into the inlet of a negative pressure air conveying system. Such a system serves as a precooler/dryer for the product while it is still sticky and hot.

The Sprout-Waldron cooker comes in two sizes. A description of the capacities, power requirements and sizes are given in Table 5. The smaller SW-450 has been used extensively as a pilot scale model and can be supplied with a wide variety of screws and larger drive to work with a greater range of products and processing conditions. Screws with higher compression ratios are available to increase the addition of mechanical energy.

The claimed advantage of the SW extruder is its ability to precook the ingredients under pressure for up to 6 min. The pressure chamber allows greater flexibility in this precooking step. The precooking leads to lower specific powers and reduced wear and maintenance costs. The extruder has been extensively for the production of pet foods, RTE breakfast cereals, pregelatinized starches, full-fat soy flour, snacks, etc.

In addition to extruders, Sprout-Waldron supplies a number of other important components for extrusion systems. These include hammer mills, coolers, pneumatic conveyors, sifters, screw conveyors, and mixer-blenders.

D. Wenger

The Wenger Manufacturing Co. (Sabetha, Kan.) is the largest manufacturer of cooking extruder equipment in the world. Their initial interests in extrusion started in

Table 5
DESCRIPTION OF SPROUT-WALDRON COOKING EXTRUDERS

Mode 1	Nominal size (in.)	Screw diameter (cm)	Screw length (cm)	Drive motor (kW)	Capacity for[a] moderate density product (MT/hr)
450	4.5	11.4	variable	56	0.91—1.36
775	8.0	20.3	157	37—112	0.91—2.73

[a]   Depends upon drive power and product.

Table 6
DESCRIPTION OF WENGER MANUFACTURING COOKING EXTRUDERS

Model	Screw diameter (cm)	Barrel segments (number)	Drive motor (kW)	Nominal capacity (MT/hr)		
				Dry snack	Pet food	Textured plant protein
X-20	8.3	3—7	22	0.07—0.11	0.36	0.07—0.14
X-25	13.3	3—7	45—56	0.11—0.41	1.36	0.16—0.32
X-155	18.4	6—8	75—93	0.7	4.5	0.70—1.1
X-175	20.9	6—10	112—150	0.8	5.5	0.90—1.4
X-200	24.1	4—8	150—186	1.14	9.1	1.8 —2.7

1946 with the construction and marketing of extrusion pelleting equipment for molasses animal feed. By 1960, extruders were being manufactured for human and pet foods. The company now maintains a very large pilot plant at their manufacturing site, where customers can try their formulations on a commercial scale before ordering equipment.

The cooking extrusion equipment manufactured by Wenger is described in Table 6. The extruders come in a great variety of configurations to match specific extrusion requirements. Scale-up between extruders is done using empirically derived "cook factors" which were designed to describe the amount of cooking occurring with various extrusion screw/barrel combinations. These factors are a function of residence time and temperature. In addition to variations in the extruder screw and barrel configurations, the extruders can be supplied with a variety of hoppers, feeders, preconditioners, and control systems. Completely automatic control panels are available to monitor and control water rate, product temperature, screw speeds, power, cutters, and all operating motors.

Wenger extruders are constructed with segmented barrels. The length of the barrel can be changed by adding sections as shown in Figure 5. These barrel segments have individual jackets and are valved so that mixtures of water and/or steam can be controlled to achieve the desired temperature in each section. The individual barrel segments are normally grooved with either straight longitudinal grooves or spiral grooves. The latter is said to give higher shear and greater cooking. On certain models, direct steam injection into the product is accomplished through steam ports into the sides of the barrel sections.

Like the barrel, the Wenger screw is also segmented giving a very large number of possible combinations. A splined shaft is used which length corresponds to the length of barrel desired. Figure 6 also shows typical screw sections along with steam locks which fit between each section. Steam locks are restrictions which increase mixing and shear in each screw segment while keeping steam from flowing back down the barrel to the feed hopper. The longer feed section of the screw is single flighted with decreas-

2 HEAD

5 HEAD

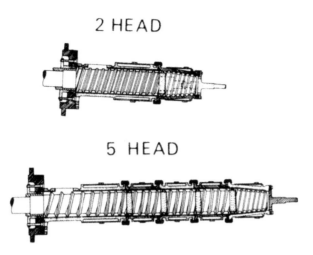

FIGURE 5. Typical construction of segmented screws and barrels in Wenger extruders. (Courtesy of Wenger Manufacturing Co., Sabetha, Kan. With permission.)

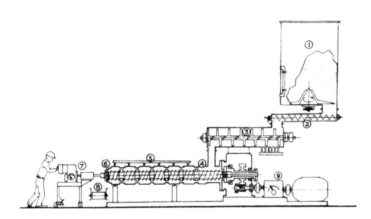

FIGURE 6. Cross-section of Wenger X-200 extruder. 1. Circular bin, 2. variable speed feed auger, 3. preconditioning chamber, 4. transition to feed section, 5. jacketed extrusion screw showing segmented screws and steam locks, 6. face cutter, 7. variable speed drive for cutter, 8. take away belt, 9. extruder drive. (Courtesy of Wenger Manufacturing Co., Sabetha, Kan. With permission.)

ing pitch. Intermediate screw sections are normally double flighted with decreasing pitch angle of approximately 12° and a trapezoidal-shaped channel. The last screw section is typically tapered to increase the compression and shear rate and achieve a rapid temperature increase and short holding times. Greater shear and product heating can be produced when screws with cut flights are used which reduces extruder conveying ability and increasing mixing. The screw can also be adjusted longitudinally in and out of the nose cone on the barrel to vary the clearances between the flights of the screw and this barrel segment. This adjustment gives further configuration variability and can partially compensate for screw and/or barrel wear.

Both high and low moisture extrusions are possible on Wenger equipment. Single- and double-shafted atmospheric preconditioners are available where water and steam can be added, and are shown in Figure 6. When the product is moistened and preheated in the preconditioner, energies associated with the extrusion are reduced.

The extruders built by Wenger can be made from a variety of materials. All human food extruders are made from stainless steel. The feeders and preconditioners use 304 and 316 SS. Barrel liners and screws are heat-treated stainless steel or stellite. Variable speed face cutters can be supplied which rotate on a central spindle on the die plate driven by an auxillary motor.

Wenger extruders have been used to manufacture an extensive variety of products including expanded snacks, pet foods, cereals, textured plant protein, animal feeds, pregelatinized starch, beverage and soup bases, and croutons. The expansion of their product capabilities has increased by coupling cold-forming extruders with cooking extruders to achieve greater forming capability and texture control.

In addition to extruders, Wenger manufactures a line of related equipment including driers, blenders, mixers, enrobers, conveyors, and slurry tanks. They claim single-supplier capability for complete extrusion systems.

E. Plastics Extruders

The use of plastics extruders for the cooking and texturizing of food products has also occurred. Now that more specialized food extruders exist, this practice is less common. No detailed discussion of plastics extruders is given here. Instead the reader is referred to Fay[1] who discusses the specification of plastics extruders for food processing applications.

VI. COLLET EXTRUDERS

Collet extruders represent a special class of cooking extruders, designed to cook and expand low moisture (<12%) degerminated corn or other cereal grits into pieces called collets. The screws on collet extruders are very short (L/D < 3:1), have pitch angles of about 6°, very high shear rates in the screw and clearance between the screw and barrel due to screw speed of about 300 rpm. For practical purposes, all energy inputs come from the dissipation of the mechanical energy so collet extruders could be called autogenous. Specially hardened screws and barrel liners are necessary to assure long production life and to reduce maintenance costs. Normally, the manufacturers have the capability to rebuild these elements once they become worn.

Some collet extruders are equipped with a jacket to maintain uniform temperatures at the die, but these play a minor role in the heat balance around the extruder. To start operation of a collet extruder, the barrel and die are often preheated with a torch.

Extensive screw and die modifications have been made to collet extruders so that they can form premoistened corn into dense collets which are fried to make an expanded product called fried corn curls. Although the internal configuration and operation of these later extruders differ extensively from dry collet extruders, they will be discussed as part of this same section since they belong to the expanded snack line of equipment.

A. Dorsey-McComb

Dorsey-McComb, Inc. (Denver, Colo.) manufactures collet extruders having a rated capacity of 135 to 450 kg/hr with drives of 22 and 55 kW, respectively. The Dorsey-McComb screw is double flighted and has shallow rounded channels which, it is claimed, eliminate the stagnant areas in the corners of channels with rectangular cross sections. The barrel has spiral grooves which are a shallow sector of a circle. Some compression is provided in the screw by decreasing channel and groove heights. The compact Dorsey-McComb extruder consists of a feed hopper, cutter, and control panel on a single machine. By a complete change of screw, barrel, and die in the collet extru-

FIGURE 7. Dorsey-McComb fried collet extruder complete with
controls and variable speed cutter. (Courtesy of Dorsey-McComb,
Inc., Denver, Colo. With permission.)

der, Dorsey-McComb supplies a fried corn curl extruder as shown in Figure 7. The
extruder adds relatively little energy to the product, forming dense collets from pre-
moisturized corn which are subsequently fried in a deep fat frier.

In addition to the standard collet extruders manufactured by Dorsey-McComb, they
recently began suppling a Series 11000 cooking/forming extruder which is shown in
Figure 8. The extruder is designed to make accurately shaped snack pieces. In the Series
11000 extruder, raw dry ingredients in the form of coarse grits are fed with a vibrator/
hopper to the long screw/barrel. The extruder has a variable speed drive for control,
and mechanical dissipation provides sufficient heat to gelatinize the cereal starches.
Using thermostatically controlled barrel zones the gelatinized dough is cooled before
being shaped in a temperature-controlled die. A capacity of 100 to 200 kg/hr is attained
with a 15-kW drive.

In addition to the collet extruders, Dorsey-McComb provides a complete line of
equipment for making expanded and coated snack foods. The equipment is constructed from stainless steel and provides continuous production capabilities for a wide
variety of expanded, shaped, and flavored snack items.

B. Krispy Kist Korn

The Par-T-Pop extruder for corn meal snacks is manufactured by Krispy Kist Korn
Machine Co. (Chicago, Ill.) The machine which makes corn twists is shown in Figure
9. A 9-kW drive is required to produce 150 kg/hr of expanded product. Different
product shapes are achieved with a changeable die plate and varying product lengths
with a variable speed fly wheel cutter.

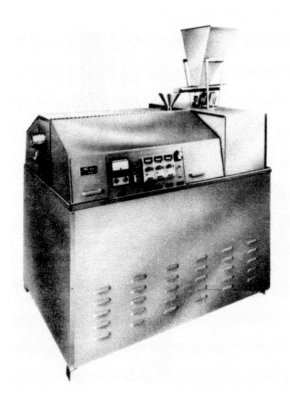

FIGURE 8. Dorsey-McComb Series 11000 cooking/forming extruder to produce precisely shaped snacks. (Courtesy of Dorsey-McComb, Inc., Denver, Colo. With permission.)

C. Manley

In the early 1960s, Manley, Inc. (Kansas City, Mo.) expanded their line of snack processing equipment, which centered around popcorn, to include collet extruders and a broader range of snack processing equipment.

Manley manufactures two basic collet extruders. The 7.6-cm diameter extruder, shown in Figure 10, has a nominal production rate of 90 kg/hr and a 10-cm diameter screw extruder will produce 135 kg/hr. Manley collet extruders have four flighted screws with rectangular channels that turn at a fixed 300 rpm. The replaceable barrel liner or sleeve is designed with rounded spiral grooves. The extruders come with a self-contained, refrigeration-cooled water recirculation system for barrel cooling and temperature control at the die. A variable speed fly knife cutter is supplied to cut products to the desired length.

The 7.6-cm diameter extruder can also be purchased with a variable speed drive, temperature measurement and a kW meter allowing for adjustments to be made for variations in the cereal grits. All extruders can produce many shapes including cylinders, balls, chips, and hollow tubes. A fried corn curl can be produced with grits moisturized to 20% followed by extrusion without expansion or significant gelatinization. The expansion and gelatinization occurs during the frying following extrusion.

In addition to extruders, Manley, Inc. can supply all the equipment necessary for the production of extruded expanded snacks including conveyors, driers, and enrobing systems.

FIGURE 9. Par-T-Pop Corn twist collet extruder.
(Courtesy of Krispy Kist Korn Machine Co., Chicago,
Ill. With permission.)

D. Wenger

Wenger Manufacturing has adapted their cooking extruder equipment for the production of expanded or fried snacks. The Wenger X-25S has been specifically adapted for the production of collet-type products at 300 to 400 kg/hr using a 45- to 55-kW drive. The smaller X-20S has a capacity of 70 to 115 kg/hr using a 22.5-kW drive. These extruders do not have preconditioning chambers but meter the grits with a vibratory feeder, along with water, directly into the throat of the extruder. The extruder can also be supplied with the horizontal mixer-feeder assembly for handling flour-type products as shown in Figure 11. Like the other manufacturers of collet extruders, Wenger can supply a complete line with product cutter, conveyors, driers, and enrobers.

To produce fried/twisted collets, Wenger has an extruder former combination with the capacity of 340 to 390 kg/hr before frying. The claimed advantage of the system is that it will handle cornmeals without premoistening or removing fines and flour. In

FIGURE 10.   Manley, Inc. Model #140 Collet extruder with 90 kg/hr nominal capacity. (Courtesy of Manley, Inc., Kansas City, Mo. With permission.)

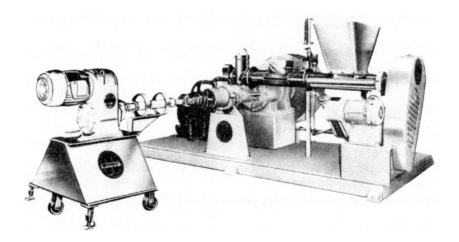

FIGURE 11.   Wenger X-25S corn curl extruder. (Courtesy of Wenger Manufacturing Co., Sabetha, Kan. With permission.)

this system, the primary extruder mixes and heats the water and corn grits and forces the material into a former consisting of a fixed serrated disc and a rotating flat disc. An adjustable clearance between these members gives the desired twisted shape to the product before frying.

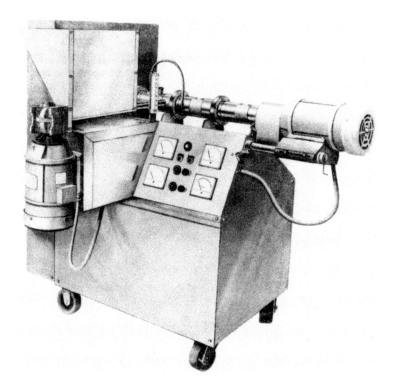

FIGURE 12. Appropriate Engineering Model 303 autogenous extruder. (Courtesy of Appropriate Engineering Manufacturing Co., Corona, Calif. With permission.)

VII. LOW-COST EXTRUDERS

Very simple autogenous extruders have been developed for the heat treatment of whole, full-fat soy beans with minimal auxillary processing equipment. The total energy input is from the dissipation of energy from their large drive motors. These extruders have no barrel jackets or preconditioning chambers. Extrusion is performed dry so that heat can be generated during the extrusion process. All machines provide some sort of variable configuration so that adjustments can be made which alter extrusion temperatures to correspond with variations in ingredients, conditions, or formulations. The applications of these simple extruders have been extended to the production of nutritious, blended food formulations of cereals and oil seeds or legumes in developing countries.

A. Appropriate Engineering

Small scale autogenous extruders have been built by Appropriate Engineering & Manufacturing Co. (Corona, Calif.) following the extruder design of the Meals for Millions Foundation. The design originally focused on a simple, small extruder to be constructed in developing countries for the production of nutritious soybean-based foods.

The AE 303 extruder, shown in Figure 12, has stainless steel construction, 22.5-kW drive with timing belt reducer, and volumetric feeders. The barrel is made up of interchangeable, unjacketed segments. The screw is built by wrapping and welding flights around a hollow shaft. Variable water application rate, screws, and dies control the operation monitored by product temperature and power consumption. The extruder

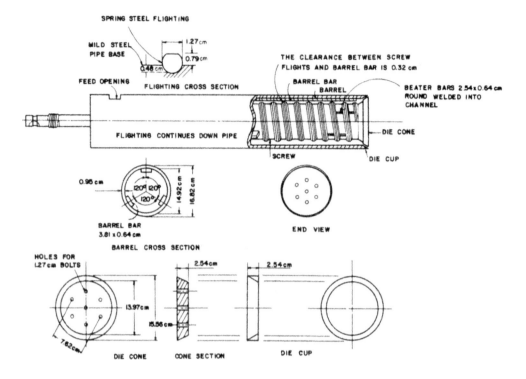

FIGURE 13.   Design of Brady extruder screw and barrel showing adjustable annular orifice at discharge. (From Sahagun, J., Thesis, Colorado State University, Fort Collins, Colo. With permission.)

can be used to expand cereals, texture plant proteins, and make a variety of consumer and pet foods.

B. Brady

The Brady Extruder, Models 206 and 2106, is manufactured by the Koehring Farm Division (Appleton, Wis.). These machines are identical but one is for portable applications while the other is placed on a stationary frame. The extruder was designed for intermittent use on a farm but with some modifications, it has been operated in a production environment on a two-shift-per-day basis.

A diagram of the Brady screw is shown in Figure 13. The screw is constructed from a hollow tube 11.4 cm in diameter, 86.4 cm in length. The flighting, in the form of a coiled spring, is wrapped onto the tube and held in position by welding. In this manner, a screw with uniform helix angle of approximately 5° and flight height of 0.8 cm is formed. To provide restriction in the screw, partial channel dams are welded between flights near the discharge called beater bars.

To control extrusion temperature, the screw is fitted with a truncated conical member at its discharge that mates with a stationary piece called a cup. Longitudinal movements of the screw can be made during operations to adjust the clearances between these parts. Such a variable annular die cannot make discreet product shapes but is restricted to the production of irregular flat expanded flakes.

The Brady Extruder can be driven at either 540 or 1000 rpm. The output is relatively insensitive to speed because of variable active length of the screw. Throughput is 550 kg/hr with a drive of 75 kW. Typical specific powers for these types of dry extrusions are about 0.10 kW-hr/kg. As is the case for extruders with high mechanical energy inputs, increased component wear should be anticipated.

FIGURE 14.   InstaPro Model 2000 autogenous extruder showing volumetric screw feeder to left of screw and auxillary fed hopper over screw. (Courtesy of Triple "F", Inc., Des Moines, Iowa. With permission.)

C. Insta-Pro

Triple "F", Inc. (Des Moines, Iowa) developed an autogenous extruder for the heat treatment of whole soybeans. The extruder is shown in Figure 14. The barrel and screw are made up of segments. Between the cast screws, that are slipped over a central shaft, are variable sized steam locks or rings to achieve increasing restrictions along the length of the screw. The barrel sections have longitudinal ribs and internal, replaceable wear rings which correspond to the location of the steam locks. Because a significant amount of energy is dissipated at the steam locks, most of the wear occurs at the replaceable hardened rings, thus reducing overall maintenance costs.

The Insta-Pro extruders come in two models. The Model 500 has a rated capacity of 275 to 450 kg/hr with a 37-kW drive. The Model 2000 has a capacity of 600 to 900 kg/hr and a 55-kW drive. Both extruders operate at 540 rpm and come with volumetric feeders, water injection systems to change the moisture of the product, and a multiple die and cutter assembly.

The applications of the Insta-Pro extruder have been in the production of full-fat soy flour, precooked animal feeds, protected protein cattle rations, and an urea-fortified cattle feed.

VIII. FORMING EXTRUDERS

Forming extruders have found application in extruding precooked doughs to produce special shapes, often requiring high pressures. To retain these shapes it is often necessary to cool the dough while it is being formed so that extensive puffing does not occur once the product leaves the die. To achieve the temperature control required for forming, these extruders are usually jacketed and often have a hollow screw for cooling water circulation.

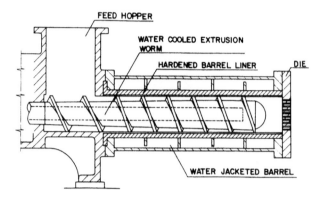

FIGURE 15. Cross-section of a Bonnot forming extruder.
(Courtesy of Bonnot Co., Kent, Ohio. With permission.)

Since forming extruders are often operated in series with a cooking extruder, variable speed drives are necessary so the output rate can be precisely matched with that of the cooking extruder.

Typical applications of forming extruders are in the formation of snack pellets which will be subsequently puffed and fried, producing a dense textured plant protein, the manufacture of RTE cereals, etc.

A. Bonnot

The Bonnot Co. makes an extensive line of forming extruders. The cross section of a Bonnot forming/cooling extruder is shown in Figure 15. The device is fully jacketed and has a hollow screw with an L/D ratio of less than 10:1. The slightly compressive screw design causes build up of pressure at the die for precise forming. The extruders are constructed entirely from stainless steel. Models having screw diameters of 5.1 cm (2 in.) to 35.6 cm (14 in.) are available with drives from 1.2 to 112 kW, respectively. Capacities can be as high as 4.5 MT/hr.

B. Demaco

The DeFrancisci Machine Corp. (Brooklyn, N.Y.) manufactures a special forming extruder Model SC-2000 for RTE cereal pellet extrusion. The forming extruder is a modification of their macaroni extruder with a small mixer and heavy duty gear box. The extruder has a rated capacity of 0.92 MT/hr.

C. Wenger

Wenger Manufacturing, Co. provides an F-20 forming extruder for low shear, low pressure forming as shown in Figure 16. The extruder has an 8.3-cm (3.25 in.) in diameter hollow screw. Heating or cooling can be accomplished by circulating steam or water through the screw or jackets.

The Wenger forming extruder has been used in conjunction with their cooking extruders to make snacks with sophisticated shapes, snack half-products, high protein cereals, dense textured meat analogs, and semimoist pet foods.

IX. MACARONI EXTRUDERS

Domestic and foreign firms supply macaroni extruders which perform similar functions but differ in certain design respects. These are discussed below by manufacturer in alphabetical order.

FIGURE 16.   Wenger F-20 forming extruder. (Courtesy of Wenger Manufacturing Co., Sabetha, Kan. With permission.)

A. Braibanti

The Dott. Ingg. M., G. Braibanti and Co. (Milan, Italy) has been a major supplier of continuous macaroni extruders for a number of years. From the beginning, their extruders have been mounted on a long-legged frame so that predriers or spreaders for long goods can be mounted directly under the presses.

A wide range of extruder capacities can be supplied by Braibanti. Their larger machines carry the name Cobra and an example is shown in Figure 17. These machines have two separate extrusion screws rather than a very large single screw, to achieve greater temperature uniformity in the dough. Cooling of the dough, occurring only at the barrel surface, and nonuniform shear rates in the channel result in temperature variations within the dough when large diameter extruders are used. These temperature variations are seen in uneven extrusion rates at the die or in product nonuniformities. To enhance extrusion rate, screws which have been Teflon® coated can be supplied.

Braibanti has developed a special semolina/water mixing system. The semolina flour is volumetrically metered continuously along with the water into a horizontal premixer (see Figure 18). The premixer runs at high speed and has a single shaft with paddles which evenly distribute the water in the flour. From the premixer, the flour falls into a twin-shafted mixer running at a much slower speed. Here, the evenly distributed water has time to be absorbed into the flour or semolina. With this system, the initially wetted particles do not lump. The slow mixing in the second mixer prevents further lumping while not interferring with the structure of the gluten.

From the mixer, the small dough pieces fall into the vacuum chamber where they are degassed. The vacuum also serves as the dough distributor in each extruder which is mounted horizontally on the frame.

B. Buhler

The first continuous macaroni presses were marketed by Buhler Brothers, (Uzwil, Switzerland). They now manufacture a wide range of new extrusion models having capacities from 600 to 3600 kg/hr. Screws with diameters 135 mm to 200 mm and L/D ratios of 7:1 cover their range of equipment. The higher capacity models have two extruders, fed by a common premixing and mixing system to achieve maximum productivity without excessive heat buildup.

FIGURE 17.   Braibanti Cobra 800 press showing twin screw construction. Product falls directly onto shaking predryer. (Courtesy of Braibanti, Milan, Italy. With permision.)

FIGURE 18.   Braibanti high speed premixer, slower twin shafted mixer and vacuum degassing chamber. (Courtesy of Braibanti, Milan, Italy. With permission.)

FIGURE 19.   Single screw Buhler TPBE-450 with 176 mm diameter screw. (Courtesy of Buhler-Miag, Inc., Minneapolis, Minn. With permission.)

Figure 19 shows a single extruder mounted on an elevated frame. The water and dry ingredient feeders are shown at the far right. The volumetric feeder operates continuously to avoid lumps in the product. Two synchronized pumps on the feeder allow simultaneous feeding of egg slurry and water. The product goes through a small single-shafted, high-speed mixer to evenly distribute the water before it falls into the large twin-shafted mixer.

A short cross-feed screw is used to connect the mixer and extruder. Between these two elements is the vacuum chamber, which can be disconnected quickly. The extrusion screw runs in a jacketed barrel. A spiral-shaped cooling jacket cast in the barrel wall enhances cooling and controls heat buildup. Small longitudinal grooves in the barrel prevent slip at the walls and increase output. The screws are designed with split flights at the discharge end to improve mixing and dough temperature uniformity. The screw can also be equipped with a bearing at the discharge to minimize screw and barrel wear.

The die assemblies of a Buhler extruder are shown in Figure 20. The quick-change hydraulic die is of the drawer type that simultaneously inserts a new die and removes the old one. Pressure and temperature gauges are located in the head for control purposes. Variable speed cutter drives and blowers, for product ventilation, are mounted directly on the short goods die head.

The Buhler extruders are fabricated from stainless steel to enhance sanitation and cleanup. All bearings are mounted outside product chambers to increase accessibility and reduce the chance for product contamination. The company calls their design simple, uncluttered, and functional.

The automated spreader for long goods was pioneered by Buhler in 1939. The machine accomodated the existing sticks in the long goods driers and the concept was quickly adopted by the industry. The device moved the sticks forward by supporting them on guide chains. The long strands of spaghetti are draped over the sticks and cut when achieving the required length. Irregular ends are trimmed off with rotary cutters and recycled to the mixer of the extruder so that minimum waste occurs.

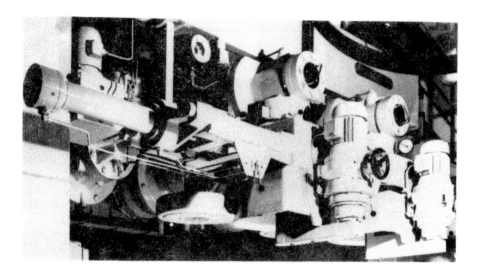

FIGURE 20. Buhler TPBD-double screw extruder with 400-mm head design. Cutter and fan drives and hydraulic die changer shown. (Courtesy of Buhler-Miag, Inc. Minneapolis, Minn. With permission.)

C. Demaco

The Demaco line of equipment is manufactured by DeFrancisci Machine Corp. (Brooklyn, New York). The macaroni extruder has been manufactured by the firm since 1941. The Demaco extruders have a single screw mounted on a frame with rated capacities of 460 to 920 kg/hr for short goods and 680 to 920 kg/hr for long goods. The firm pioneered in the design and construction of long goods spreaders.

One of the claimed advantages of the Demaco extruder is its simplicity and sanitary design. The machine is constructed of stainless steel and easily disassembled for cleaning. All bearings are mounted outside the equipment to increase accessibility and remove the chance for grease to enter the product. The design avoids inaccessible crevices which are difficult to clean and, therefore, are a sanitation problem. The semolina and water feeder is mounted directly above the twin-shafted paddle mixer. The entire mixer is maintained under vacuum which eliminates the need for a transfer auger to the extrusion screw, again aiding clean-up.

The Demaco S-2000, shown in Figure 21, has a single extrusion screw and can have twin heads to achieve 920 kg/hr capacity. This design allows a slower extrusion rate through the dies to maintain product integrity for intricately shaped short goods, improved product color, and surface appearance.

The application of macaroni extruders to specialized can-filling equipment has been an innovation of Demaco. With their equipment, wet spaghetti is accurately metered into No. 300 sized cans at the rate of 200 cans per minute with the filler shown in Figure 22. The spaghetti is extruded to a proper length and cut so it falls directly into the waiting can. Tomato sauce and other ingredients are metered on top of the spaghetti in a separate filling operation and a retort cooks and mixes the components to achieve cooked canned spaghetti.

D. Malard

The Societí André Malard et Cie, (Montreuil, France) manufactures pasta presses. An example of their AMC 600 press is shown in Figure 23. This machine has a rated capacity through a single screw of 600 kg/hr.

The extruder, constructed from stainless steel or chrome plated, resists corrosion.

FIGURE 21. Demaco Model S-2000 short goods extruder having a single extrusion screw and head. (Courtesy of De Francisci Machine Corp., Brooklyn, N.Y. With permission.)

FIGURE 22. Automatic Demaco DC-1500 direct can-filling spreader for spaghetti. (Courtesy of De Francisci Machine Corp., Brooklyn, N.Y. With permission.)

A variable speed volumetric feeder is controlled electronically to achieve a constant feed rate. A constant level water wheel metering device is used to maintain an independent water feed.

A conventional, twin-shafted paddle blade mixer is used to blend the semolina and water. A transfer auger serves as a seal between the mixer and screw. The vacuum system, attached directly to the barrel near the feed port, degasses the dough.

The extrusion screw is chrome plated and hardened to resist wear. A jacketed barrel has circulating cooling water to maintain a low dough temperature and prevent heat buildup. The surface of the barrel is grooved to prevent slip.

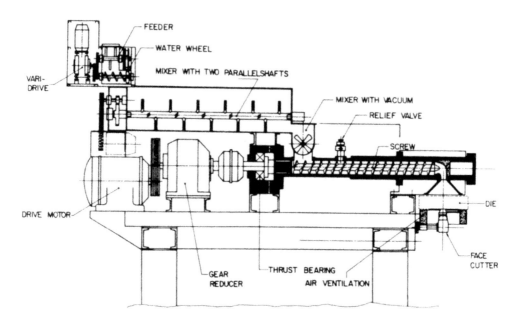

FIGURE 23.   AMC 600 pasta extruder. (Courtesy of Malard et CIE, Montreuil, France. With permission.)

An electronically controlled variable speed drive is coupled to a vertical face cutter on the die. Ventilation around the die causes surface drying of the pasta and prevents sticking.

A separate control cabinet for the extruder has a number of built-in safety features to warn of overloads and automatically shut down equipment should they occur.

E. Pavan

Pavan SPA, (Padua, Italy) manufactures and supplies a complete line of pasta equipment. They manufacture four sizes of single-screw pasta extruders with rated capacity ranging from 300 to 700 kg/hr. Higher capacities are achieved by mounting multiple extruders on a single stand.

The Pavan extruder has a patented flour/water metering unit which is suitable for vacuum mixers. The flour is fed through rotary valves and the water metering device uses the constant level principle. Once metered, the flour/water mixture goes through a high speed centrifugal mixer before it enters the vacuum mixer. The flour/water metering units and centrifugal mixer are shown above the horizontal mixer in Figure 24.

The dough mixer has a single shaft with blades of stainless steel. The entire mixer is under vacuum which is claimed to give a pasta with superior color and appearance.

In 1947, Pavan introduced extrusion screws with grooved barrels to reduce slippage at the wall, increase output, and pressure. The barrels are made of cast iron and the screws have a chromium plating.

A number of safety features are incorporated in the Pavan extruders. A pressure indicator is mounted on each head to warn of high pressures resulting from poor mixing. Shear bolts are also used in the head so that when pressures exceed 14 MPa, the dough is immediately released without damage to dies, screws, or the drive train. Watt meters are also supplied as a safety precaution and aid in uniform production.

The die heads on a Pavan short-good lines are shown in Figure 25. The large blowers are clearly visible to accommodate the higher extrusion speeds. The air is also heated to hasten surface drying and to maintain flexibility for both long goods and nested

153

FIGURE 24. Single Pavan P7 extruder (700 kg/hr) attached to a long-goods spreader. Press is mounted over pre-dryer which is attached directly to final dryer. (Courtesy of Pavan SPA, Venetia, Padua, Italy. With permission.)

FIGURE 25. Pavan P10 press with two extruders mounted over shaker pre-dryer. (Courtesy of Pavan SPA, Venetia, Padua, Italy. With permission.)

products. Efforts have been made to reduce the length of the transition between the extruder screw and die assembly to minimize back pressure and improve uniformity of flow.

X. LABORATORY EXTRUDERS

With the increased interest in food extrusion, there has been a need for small-scale laboratory extruders that could be used for product development, fundamental extrusion studies, and ingredient testing. The difficulty with small extruders involves ingredient feeding, a relatively larger portion of heat entering the product through the jackets, unknown scale-up relationships between small and large extruders, and measurement difficulties. Work is underway which has addressed these issues so that laboratory extruders are finding an increasingly important role in the research laboratory.

Small-scale, pasta-forming extruders are available from Buhler-Miag (Minneapolis, Minn.) and Demaco. Wenger provides the X-5, a scaled-down version of their cooking extruder having multiple barrel segments and varying length screws. Several food laboratories have used small plastics extruders for research work. Examples of such equipment are those manufactured by Akron Extruders, Inc. (Canal Fulton, Ohio) and Killion Extruders, Inc. (Pompano Beach, Fla.).

C. W. Brabender Instruments, Inc. (South Hackensack, N.J.) manufactures laboratory extruders having 10:1 and 20:1 L/D ratios with diameters of 1.9 and 3.2 cm; an example is shown in Figure 26. The extruders are driven by an SCR-controlled, DC motor giving widely variable speed control. Continuous torque measurements are made with the Plasti-Corder® along with a dial indicator for screw speed. The barrel can be heated or cooled, and controllers and recorders are provided. Pressure measurements can be made with a variety of dies available for purchase. Timbers et al.[7] describe modifications to the feeding system and process control of the extruder to add to its versatility.

XI. COOKING EXTRUDERS (TWIN SCREW)

This book focuses primarily on single-screw extrusion but some mention should be made of twin-screw extruders that have been recently introduced into the food extruder market. Twin-screw extruders have two parallel screws which can come in different configurations, as shown in Figure 27, described as intermeshing and nonmeshing. The nonmeshing or corotating screws act similarly to two single screw extruders. These machines have good mixing action and transport of materials. However, one channel is less filled than the other which can result in uneven flow out a multiple-holed die and steam can flow back through the relatively open channel, interrupting operation.

The intermeshing counter-rotating models provide C-shaped chambers between the flights which mix and convey the product. The flow developed by the intermeshing screws approaches that of a positive displacement pump. Because some leakage does occur, the speed of the screws can be adjusted to achieve constant outputs against various back pressures. The chambers formed between the screws are completely filled with material to avoid metal-to-metal contact and reduce wear. The speeds of the screws are slow and the shear rate is low so a significant amount of energy must be transferred to the product through jackets around the barrel. The theoretical aspects of twin-screw extrusion are given in Janssen.[4]

The twin-screw extruders are obviously complex and more expensive than single-screw extruders, requiring closer tolerances and a complex drive and thrust bearing arrangement. Despite these disadvantages, they are able to perform extrusions under very low moisture conditions which offer certain operating advantages because product drying may not be required. In some instances, overworking of the product may be a disadvantage of the twin-screw design.

FIGURE 26. Brabender Plasti-Corder® laboratory extruder shown with instrumentation. (Courtesy of C. W. Brabender Instruments, Inc., South Hackensack, N.J. With permission.)

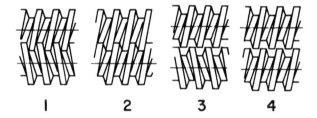

FIGURE 27. Different kinds of twin-screw extruders. 1. Counter-rotating, intermeshing, 2. co-rotating, intermeshing, 3. counter-rotating, non-intermeshing, 4. co-rotating, non-intermeshing. (From Janssen, L. P. B., *Twin Screw Extrusion*, Elsevier, Amsterdam, 1978. With permission.)

A. Continua®

The Werner-Pfleiderer Corp. (Waldwick, N.J.) is now offering intermeshing twin-screw extruders for food processing applications. The entire barrel assembly is on a track so that it can be easily removed for cleaning and inspection of the screw as shown in Figure 28. Such a design is necessary because large capacity machines have very heavy barrels. Screw elements on the Werner-Pfleiderer can consist of kneading blades as well as intermeshing screws having channels with rounded bottoms. Such an arrangement provides flexibility of machine configuration. With the intermeshing twin-screw extruder design, the company can configure extruders which meet a wide range of shear, conveyance, pressure buildup, temperature, and residence time requirements for varying food products.

The Continua® line of extruders is available to cover extrusion rates of 0.4 to 21 MT/hr with drive powers of 16 to 870 kW maximum. The barrels of the extruders are jacketed so as to accept liquid or vapor media. Vents can be used in intermediate barrel sections to relieve moisture part way through the extrusion process.

FIGURE 28.   Continua® twin-screw extruder shown with barrel and kneading screw. (Courtesy of Warner-Pfleiderer Corp., Waldwick, N.J. With permission.)

B. Creusot-Loire

The Creusot-Loire extruder is manufactured in France and distributed by Pembertons Food Processing Equipment, Inc., (Vermilion, Ohio). The equipment comes in a variety of sizes to achieve a range of production capacities ranging from 0.03 to 2.0 MT/hr. The corresponding drive motors for these extruders are 7.5 and 260 kW, respectively. Typical maximum specific energy is 0.10 kW-hr/kg.

The Creusot-Loire extruder has twin intermeshing screws. A picture of the machine is shown in Figure 29. The extruder has a conveyor for the barrel which makes easy access to the screw. To add additional heat, an induction heating system is employed around the barrel which is said to achieve very precise cooking temperatures.

Extruding at very low ingredient moistures, the Creusot-Loire has been successfully used to puff whole grains, precook vegetables and flours, texturize plant proteins, gelatinize starches, and expand pet foods.

C. M-P Continuous Mixers

Baker-Perkins, Inc. (Chemical Machine Div., Saginaw, Mich.) offers twin screw continuous mixers having corotating screws of variable size and characteristics. The "clam shell" split and hinged barrel allows total exposure of the screws and facilitates cleaning. The screws have three distinct zones: feed, mixing, and discharge. The rotating members on the mixing shaft are lens-shaped (see Figure 30), which provides control of mixing, heat transfer, and throughput rates. To further control mixing, the M-P mixers have a variable orifice within the barrel where clearances are adjusted by small axial movements of the screw. The variable orifice controls back pressure and resulting mixing intensity.

To aid heating or cooling, the machine has jackets on the barrel and cored screws.

157

FIGURE 29. Creusot-Loire Model BC 45 twin-screw food extruder with 15-kw DC drive. (Courtesy of Pembertons, Food Processing Equipment, Inc., Vermilion, Ohio. With permission.)

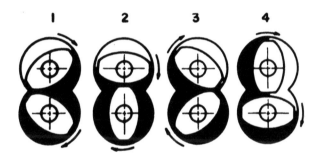

FIGURE 30. Lens-shaped mixing elements used with M-P twin screw machines. (Courtesy of Baker-Perkins, Saginaw, Mich. With permission.)

Models come in sizes having capacities of 0.35 to 11.5 MT/hr with drive motors of 55 to 1800 kW. The mixer-extruder has been used for processing edible proteins, pet foods, gum bases, and other dough-like materials.

D. Textruder®

Twin-screw extruders with conical-shaped screws are manufactured and distributed by Textruder Engineering Ag. (Zug, Switzerland) and are shown in Figure 31. The extruders come in 6 sizes from 0.2 to 4.0 MT/hr. The counter-rotating screws are driven by a variable speed motor in the region of 5 to 10 rpm which gives long wear for hardened barrels and screws. The conical shape of the screws is claimed to have advantages in feeding flour-like materials under dry extrusion conditions providing more area for thrust bearings and creating an internal compression. The texturing of soy flour having low NSI (nitrogen solubility index) is claimed with the design.

Little heat is added in the extruder from the mechanical dissipation of the mechanical energy. Most heat comes from oil circulating in the jackets and cores of the screws. The high level of mixing within the screws allows very uniform temperature so that

FIGURE 31. Textruder® twin conical screw extruder having 200-kg/hr capacity.
(Courtesy of Textruder Engineering AG, Zug, Switzerland. With permission.)

biological materials are not burned or otherwise damaged. Also, the Textruder® has
been used to make dry expanded biscuits from cereal flours having an accurate and
reproducible shape.

XII. DESIGN OF FOOD EXTRUDERS

The design and construction of food extruders has evolved considerably over the
past 30 years. In addition to improved materials of construction, greater emphasis has
been placed on the sanitary design, ease of maintenance, and steady operational performance of food extruders. Williams et al.[8] and Hoskins[3] have discussed a number
of these design improvements which are summarized below.

A. Materials of Construction
The metal surfaces in contact with food materials should be noncorrosive and non-
toxic which can be hardened for components where wear occurs. The use of stainless
steel in mixers and blenders has been common for a number of years but now heat-
treatable stainless steel (400 series) is being used extensively for the construction of
screws and barrel liners.

Many food extruders used on low moisture products are not stainless steel because
contact with moisture is minimal and stainless steel is more expensive and less easily
hardened. Some extruders are being used to form filled products such as ravioli and
egg rolls with meat emulsions or other stuffing materials. In these cases, wet cleanup
procedures are required and stainless steel construction of all contact surfaces with the
food is necessary.

Chrome plating has been used as a method for increasing the corrosion resistance
of extruder parts. Although the resulting parts are very smooth, the chrome plating
has a tendency to flake off with time. These flakes can contaminate the food products.

B. Sanitary Design
The sanitary design of extruders involves many things. First, all surfaces must be
smooth and free from crevices where food can accumulate and which are difficult to
clean. Special care needs to be taken at shafts and bearings so that lubricants will not
contact or contaminate food materials or conversely, food will not contaminate lubricants.

Ease of disassembly of the food extruder is essential so that all surfaces contacting the food can be exposed, cleaned, and inspected. Blenders should have large removable covers so all parts are exposed. New barrel designs which are slit and hinged allow opening of these members without damage to the heavy, bulky pieces.

Many times extruders are cleaned by washing. In these cases, all parts must fully drain so that no water is left standing in the equipment. Wash water left standing contributes to microbial growth and contamination of fresh food products that enter the equipment.

The motor control center, extruder instrumentation, and other controls are usually mounted in a dust- and water-tight, electrical control cabinet separate from the extruder. This cabinet can be covered during cleanup operations if excessive water splashing occurs.

The stand supporting the extruder and related equipment should be constructed out of tubing rather than angles to further eliminate harborages for food and ease cleanup. Accessibility to all parts of the extruder is essential if proper cleaning is to be performed.

C. Reduced Wear

Contact between the screw and barrel cannot be avoided and wear will occur with time. Often the barrel is fitted with internal sleeves which can be replaced when wear occurs. Screws can be rebuilt by adding metal to the worn flights and grinding to the required shape. To reduce contact, a bearing at the discharge of the screw has been used with some extruders.

D. Steady Performance

Uniform operation is a function of extruder design and the maintenance of steady feed and extrusion conditions. The extruder must be capable of handling feed materials without bridging or blockage, while maintaining a uniform pressure behind the die. Accurate feeders and flow controllers are necessary to assure uniformity of feed composition and rate.

E. Quality Cutter

Clean cutting at the die surface is necessary to achieve neat, uniform extruded pieces. Thin knives tend to cut cleaner than thick knives with a pitched cutting surface. The cutter must be precisely placed relative to the die face and have a variable speed capability to match the extrusion output.

XIII. AUXILLARY EXTRUSION EQUIPMENT

A large number of pieces of equipment are necessary to support the operation of a large continuous food extruder. Extensive bulk ingredient storage and handling equipment and blending and mixing equipment are required. Discussions of these devices are beyond the scope of this text; many texts and references are available on the subject.

Some postextrusion equipment such as cutters and driers will be discussed briefly to provide some familiarity with the equipment and techniques used in these areas. A discussion of enrobing equipment is given in Chapter 9, which discusses flavor and color applications to extruded products.

A. Cutters

Cutting of the hot, sticky food materials is a difficult problem. Basically three types

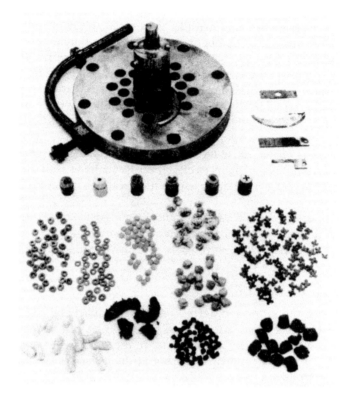

FIGURE 32.   Centrally mounted face cutter. Cutter knife options,
die inserts and shaped products are shown. (Courtesy of Anderson-
Ibec, Strongsville, Ohio. With permission.)

of cutters are employed at the extruder die. One is a high-speed, rotary cutter mounted centrally on the face plate of the extruder die. An example of such a cutter is shown in Figure 32, along with some various knives and shaped products produced. (A second basic type of cutter is the fly knife with its axis of rotation to the side of the extrusion barrel.) Such a cutter, shown in Figure 33, offers much higher knife velocities passing through the product than the centrally mounted rotary cutter. Lastly, guillotine cutters have been designed and used when relatively long pieces or slow extrusions are required.

The adjustment between the cutting knives and the face of the die is usually done before start-up. Knives are often attached to the rotating cutting member with machine bolts allowing for their individual positioning. Some cutters have been designed so the cutting head can be moved relative to the die face while in operation. Extreme care must be exercised here to assure that damage is not done to the cutter because of negative clearances. Other cutters use flexible, spring steel knives that rest against the face of the die and maintain constant contact.

Paladini[5] has given a clear discussion of considerations that must be made in designing the knives or cutters for extruders. The angle of attack is shown in Figure 34. The diagrams on Part 2 of this figure show a blade with no angle of attack, that merely chops the extrudate, while those having a high angle of attack have a sliding motion which enhances clean cutting. The importance of the sliding motion in cutting can be illustrated by the value of moving a knife back and forth while slicing bread rather than merely pushing the knife directly through the product. A curved blade in Part 3 of Figure 34 illustrates a knife with a variable angle of attack. The higher the angle of attack, the smaller driving force required to push the knife through the product.

FIGURE 33.   Fly knife cutter in housing. Note axis of rotation of the cutter is to the side of the longitudinal axis of extrusion screw. (Courtesy of Bonnot Co., Kent, Ohio. With permission.)

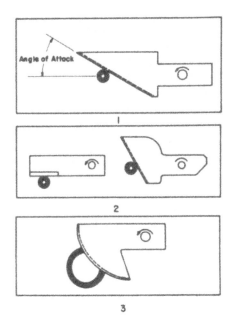

FIGURE 34.   Cutter blade angle of attack.
(From Paladini, R., *Plastics Technol.*, 23(5), 83, 1977. With permission.)

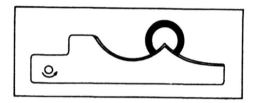

FIGURE 35. Penetrating knife cutting hollow tube. (From Paladini, R., *Plastics Technol.*, 23(5), 83, 1977. With permission.)

To reduce cutting force, it is important that the knife be as thin as possible, consistent with mechanical strength to prevent breakage, vibration, and fatigue, and reduce cutting and impact forces. Vibrations of the knife can be a serious problem because they result in wavy cuts. The knife width should be kept small so that the trailing edge does not interfere with the continuing flow of the extrudate. To partially overcome this problem, knives are usually pitched at an angle of about 5° to the die face so the trailing edge does not smear the product coming from the die hole.

Knife sharpening is more of an art than science. Knives with a long bevel become dull quickly and require resharpening frequently. Conversely, knives that are relatively blunt require more cutting force and usually produce fine pieces at the cut called dust.

Cutting hollow tubes can be a problem because they tend to collapse when the knife reaches them. A penetrating knife illustrated in Figure 35 is one solution to overcome the problem.

B. Driers/Coolers

Proper drying and cooling of extruded products is an essential portion of the process. In macaroni products, drying must be done slowly under high humidity conditions to achieve low moistures in the product without the formation of stress cracks.

Many times, drying is done in several steps. The first drying is usually called predrying, where the surface moisture of the product is removed to prevent sticking or clumping of the product. Oscillating tray driers are frequently used here for they keep the product moving while it is still sticky. In cases where pneumatic takeaway systems are employed directly after the die, predrying can occur in the pneumatic system which may employ heated air.

Driers with perforated belts are by far the most commonly used. An example is shown in Figure 36. There are normally several belts within the dryer having independent speed adjustment so the product can be piled at varying depths as the drying process proceeds. Air flow, temperature, humidity, and residence time can all be controlled to achieve the desired drying effect.

Rotary drums have been used as product driers for collets and similar products. The drums, fitted with lifting flights, tumble the product through heated air or are exposed to radiant heat from heaters mounted inside the equipment.

Coolers normally follow the driers to bring the product down to ambient temperature after drying, thereby preventing "sweating" when it is piled before packaging. Coolers usually have perforated belts with ambient or chilled air blown through the product.

FIGURE 36. Belt-type dryer used for drying extruded product. (Courtesy of Procter-Swartz, SCM Corp., Philadelphia, Pa. With permission.)

REFERENCES

1. **Fay, N. V.,** How to specify a food extruder, *Food Eng.,* 46(11), 91, 1974.
2. **Harper, J. M.,** LEC technology: where does it fit?, in *Low-Cost Extrusion Cookers,* LEC-5, Wilson, D. and Tribelhorn, R. E., Ed., Colorado State University, Fort Collins, 15, 1979.
3. **Hoskins, C. M.,** Sanitary extruder is wet-cleanable, *Food Eng.,* 44(11), 82, 1972.
4. **Janssen, P. B. M.,** *Twin Screw Extrusion,* Elsevier, Amsterdam, 1978, 30.
5. **Paladini, R.,** Extrusion cutter knife design: six basic pointers, *Plastics, Technol.,* 23(5), 83, 1977.
6. **Rossen, J. L. and Miller, R. C.,** Food extrusion, *Food Technol. (Chicago),* 27(8), 46, 1973.
7. **Timbers, G. E., Paton, D., and Voisey, P. W.,** Modifications to improve the operational efficiency, process control and versatility of a Brabender laboratory extruder, *Can. Inst. Food Sci. Technol. J.,* 9(4), 232, 1976.
8. **Williams, M. A., Horn, R. E., and Rugala, R. P.,** Extrusion — an indepth look at a versatile process. II, *Food Eng.,* 49(10), 87, 1977.

Chapter 8

EXTRUSION OPERATIONS

I. INTRODUCTION

Extruders are complex devices which cause significant changes in food ingredients under conditions of high temperature, shear, and pressure maintained for only short periods of time. The operation of an extruder is, however, relatively straight forward, once the operating principles are understood. This, coupled with the high productivity of extrusion equipment has resulted in its acceptance as the principle processing tool for a number of food process operations.

This chapter will emphasize various aspects of extruder operational technology. The general description of the start-up, steady state operation and shut down of extruders will be covered first. Control of extrusion processes will follow stressing the importance of certain variables and their manual and feedback control. Maintenance problems and procedures related to extrusion equipment will also be discussed. Finally, operating costs for extrusion systems and the variation of energy requirements and costs with differing extrusion moistures will be examined.

II. OPERATIONS

The following is an overview describing the general operations of a food extruder. Each specific extruder has its own idiosyncracies, which require a specialized understanding of the operating instructions provided by the manufacturer. In this discussion, the general sequence of steps associated with the start-up, steady state operations, and shutdown will be enumerated and discussed.

A. Assembly

Before start-up can occur it is necessary to assemble the extruder and related equipment. The amount and type of assembly required will depend upon the particular extruder used and the cleanup procedures employed after previous runs.

Some food extruders are designed so that the screw and barrel are made in segments which must be assembled to achieve the correct operating configuration. Normally, each extruded product has its own configuration. To achieve proper assembly, marked parts are laid out systematically so that assembly naturally gives the proper order. During assembly, all component parts are visually inspected for wear or defect which may cause an operational abnormality. Obviously, those requiring replacement are replaced during the assembly process.

As part of assembly, it is necessary to assure that all fasteners are tight and parts are aligned. Mating surfaces must be clean to assure parts fit together properly. If retaining bolts or nuts have a torque specification, it should be adhered to rigidly. Make sure all bolts are of proper size and the threads meet the strength (hardness) specification for their intended duty. Finally, water and/or steam connections to barrel are attached as required.

Extruders which have a one-piece screw and barrel are much easier to assemble. In these cases, the barrel often remains part of the extruder base or stand and is only removed when extensive maintenance is performed. After assembling the barrel and screw, it should always be turned by hand to assure it moves easily without any binding or restriction.

Some extruders have movable screws, relative to the thrust bearing, which allow adjustments of the clearance between the conical screw section at the discharge and the mating barrel section. Adjustments are normally made cold and a feeler gauge is used to set the clearance.

To complete the screw assembly, the die plate is bolted to the last section of the barrel. In certain cases, spacers or breaker plates are placed between the die plate and the barrel to give some extra volume at the screw discharge end to improve pressure uniformally behind the die plate and increase the residence time of the food in the extruder. If a die plate with inserts is used, the inserts are usually placed into the plate before it is mounted making sure that they do not protrude from the discharge side.

Assembling and readying the cutter always preceeds start-up. Freshly sharpened blades are first inserted into the rotating cutting knife assembly. Adjustment of the knives relative to the die head is critical for extended life of the blades and the production of a cleanly cut product. Usually, a very small clearance is maintained between the die face and blades (0.05 to 0.2 mm) when the system is cold. For cutters that provide adjustment of the knives relative to the die face during operation, it is critical that all blades have exactly the same relative clearance to the face at the start of any run.

Prior to start-up, all feeders and regulators need to be checked to determine if they are operating properly. In some cases, feeders need to be calibrated between runs to maintain standard operating conditions and control. All steam lines should be blown down to remove condensate or any accumulation of undesired materials. Magnets on dry ingredient feeders need to be cleaned to prevent the possibility of tramp metal from entering the extrusion screw. Finally, all guards and safety devices should be in place before start-up begins.

B. Start-Up

The purpose of the start-up procedure is to bring the extruder to operating condition and equilibrium as quickly as possible. If start-up is a prolonged process, the cost of the raw ingredients required for this period is substantial. In certain cases, some of the off specification product made during start-up can be reprocessed (reground) so the noningredient processing cost only contributes to start-up costs. Developing a specific routine to start a food extruder is very necessary to assure attaining uniform operations rapidly, minimize the loss of production time and wasted product, and to avoid any mechanical breakdowns or physical damage to the extruder and related equipment.

Prior to feeding ingredients to the extruder, the extrusion system should be brought as closely as possible to the steady state operating temperature. For cooking extruders, this usually means the introduction of steam into the jackets surrounding the barrel or to the barrel itself and into the preconditioning chamber, if one is used. On some autogenous extruders with no jackets, the barrel and die may be preheated with a torch to reduce start-up time. The total mass of the preconditioner and extruder is very substantial so that preheating is essential to compensate for the heat capacity of these members if shortened start-up times are to be achieved.

During the preheating step, the operator rechecks the dry ingredient, slurry, and water feeders or controllers to see if they are operating properly. It is common practice to start these feeders in a diverted position and to take a manual feed check by catching a timed sample and weighing it. Initially, the dry ingredient feeder may be set below the required feed rate with a high moisture input so the start-up is begun at reduced rates and high moisture conditions. Before the feed is directed to the extruder, the preconditioner and extruder motors are started. Rapid wear of the extrusion screw and barrel will result if they are operated without being filled with food materials, so the close timing of these start-up procedures is necessary.

Before the food enters the extruder, it is important that all down-stream equipment be started and operating. This includes takeaway belts or systems, cutters, dryers, coolers, enrobers, elevators, etc. Because driers take some time to warm up, their start-up may proceed any of the actual start-up operations on the extruder itself.

The initial reaction of the extruder, as feed enters, is for the screw to convey the feed forward, completely filling the voids in the channel. The operator watches the power draw of the motor and the condition of the extrudate. Once it is clear that the screw is full and wet extrudate is exiting the die holes, the feed rate is slowly increased in small steps at 1 to 2 min intervals. The water/slurry rate is simultaneously adjusted so the product becomes progressively drier, approaching the desired final moisture content. If steam injection into the product is used to provide part of the energy requirements, the steam valve(s) are slowly opened during the start-up phase. In these cases, the water will be reduced rather quickly as the steam is increased, recognizing the condensing steam will add a significant quantity of water to the product.

The power being drawn by the extruder drive, the die temperature, and sometimes pressure of the extrudate measured at the die and the condition of the extruded product are the key parameters observed by the operator signalling the completion of the start-up period. Initially all these parameters will move quickly toward the desired equilibrium conditions in response to changes in the operating parameters made by the operator. But the final adjustments necessary to reach equilibrium may take an hour or more. A good operator can normally bring a larger extruder online, making acceptable product, within 10 to 20 min after feed is introduced to the extruder.

Extra attention has to be paid in that period following start-up until true equilibrium is reached. Equilibrium occurs when the source of the energy inputs (mechanical dissipation, heat transfer through jackets, and direct steam injection) become balanced to achieve the required extrusion temperature and cooking environment. The mass of the extruder screw, barrel, and other components is very large compared to the mass of the product in the machine at any one time, so achieving thermal equilibrium is a correspondingly slow process.

The final adjustments made to an extruder represent fine tuning of the moisture added, the temperature of the barrel jackets or extrusion screw, the amount and input location for steam to the preconditioner and/or barrel, the dry ingredient feed rate, and, in some cases, the speed of the extrusion screw. The operator needs to be taught which of these adjustments should be made to achieve the desired results in the shortest period of time. The condition (appearance, texture, taste, etc.) of the extruded product is usually the final indication that equilibrium operation and required product specification have been achieved. Minor adjustments to the cutter speed will also be required to produce product shapes and lengths within specifications.

C. Steady State

Once an extruder is "lined out" and operating at "steady state," very few minor adjustments should be required. When an extruder is operated manually, long term drifts in feed, water, and steam rates will have to be compensated for by adjustments of these variables. More importantly, variability in raw ingredients or extruder wear can cause shifts in operations with time. No general rules can be given because the nature of these changes can be complex. Some general statements concerning trouble shooting are given later; most of these will apply, however, to adjustments which are made to compensate for variations in extrusion operations.

To assist the extruder operator in maintaining steady state conditions, it is common

EXTRUDER _____ PRODUCT _____

DATE _____ OPERATOR _____ SHIFT _____

PROCESS VARIABLE	STD	TIME								
1 FEED RATE, kg / min										
2 WATER RATE, l / min										
3 STEAM RATE										
a. PRECONDITIONER, kg / min										
b. BARREL, kg / min										
4 SCREW SPEED, rpm										
5 PRECONDITIONER SPEED, rpm										
6 PRECONDITIONER TEMP., °C										
7 POWER, kW										
8 JACKET TEMP										
a. #1, °C										
b. #2, °C										
c. #3, °C										
9 KNIFE BLADES										
PRODUCT VARIABLE										
1 TEMPERATURE AT DIE, °C										
2 PRESSURE AT DIE, bar										
3 MOISTURE, %										
4 COLOR, APPEARANCE										
5 BULK DENSITY, g /cm^3										
6 PRODUCT SIZE, L x W, mm										
7 RATE CHECK, kg / min										
COMMENTS:										

FIGURE 1.   Typical extrusion data log sheets for periodic recording of critical operating variables.

practice to have him log the data on critical variables affecting the operation at least once an hour. The variables appearing on log sheets differ with the operations but consist of a combination of independent and dependent extrusion variables given in Chapter 6. A typical log sheet is shown in Figure 1 listing all important variables. The column marked standard on the log sheet gives the range of values of the variables listed occurring under normal or standard operating conditions. Deviations from the standard should not continue for extended periods of time without corrective action or an explanation under the section for comments.

One additional purpose for the manual data logging by the operator is to emphasize the necessity of checking and maintaining all extrusion variables within specification. Otherwise, the extruder may run for long periods of time before problems or deviations are noted. The continual checking of variables with small corrections to maintain standard operating conditions usually helps avoid "sudden changes from specifications." When corrections in processing conditions are noted by the operator, they should be made slowly, using small incremental adjustments of 5- to 10-min duration. Rapid adjustments in variables tend to upset operations and/or cause more serious operational difficulties.

D. Trouble Shooting

The extruder operator typically has direct control of feed rate, water rate, steam addition to the preconditioner or screw, and temperature of the jackets. It is with adjustment of these independent variables that product temperature, die pressure, power, condition of the product, etc. are controlled. An experienced, qualified extruder operator has either been taught or acquired the knowledge or skill allowing him/her to adjust the extruder and achieve the desired product.

Many extrusion problems occur because of variations in input rates of feed, water, or steam which, directly relate to the design of the system. For example, the level of material in a hopper above a volumetric feeder can change its output. Some feed materials tend to segregate with handling so that variations in the granulation of the ingredients will change with time. Temporary blockage of the feed hopper or feed transitions will rapidly alter the feed rate. Under no circumstances should this blockage be cleared with a metal rod or pipe which would severely damage the screw if it gets into the flights. Wooden or polyethylene rods only should be used. All these changes can cause fluctuations in feeder performance and the way the extruder behaves.

Unless care is taken in the design of water and steam supply lines, pressure in these lines can fluctuate severely as the demand for these items vary in other portions of the plant. These fluctuations result in changes in the input rates of these items and markedly alter extruder operations. The elimination of these problems, which plague most extrusion operations, is through initial care in the design of the system or expensive redesign and alteration of existing systems so fluctuations do not occur. Pressure regulator controllers on all lines is one necessary step toward preventing some of the fluctuation problems but they cannot compensate for a poorly designed system with undersized supply lines.

Changes in the power consumed by the extruder drive is a very sensitive indicator of extruder operations. When an increase in kWs is seen, several corrective actions are suggested. Either the feed rate should be reduced slightly or the water rate increased slightly. The resulting effect is to have a wetter extrudate. Equation 4.66 shows power decreasing with viscosity which is reduced by increases in moisture. The choice of variable to be changed is rather arbitrary, but it should be the one which is controlled most easily and accurately. Along with increases in power consumption, increases in product temperature and head pressure are also frequently observed.

If increased product temperatures are desired but motor power is within acceptable limits, jacket temperatures may be increased or more steam can be added to the product with a corresponding reduction in water rate. Increases in product temperature or moisture normally result in somewhat lower head pressures at a constant flow rate. The ΔP at the die is directly related to flow and inversely proportional to viscosity as shown by Equation 4.27.

The extruded product characteristics may change because of ingredient variations, extruder wear, or other unexplained variations. Commonly, these changes are in taste, texture, density, etc. To increase a desired cooked flavor, product expansion, and produce a softer texture, the extrusion temperature can be increased and/or the extrudate moisture reduced.

Irregular-shaped product can occur for many reasons. The cause may be at the beginning of the process where raw ingredients and moisture are not dispersed properly to give a homogeneous mixture. Changing the blending procedure or method of adding water to the extruder may solve the problem. More likely, however, irregularly shaped produce signifies surging, plugging of the breaker plate or die, or improper cutter knife operation.

The possible causes and corrections for surging are discussed below but its presence results in uneven flow through the die and irregular product. Plugging of the breaker plate or die is usually evidenced by reduced flow through one or more die holes. It is normally very difficult to clear the cause of the blockage without physically removing the die and breaker plate, cleaning, and then replacing them. Some extruders have hydraulically operated quick change die and breaker plates which allow for their change with only momentary stoppage of the extruder operation. Cutter problems typically require clearance adjustments or replacement of knives.

Surging represents unsteady state extruder operations which are often correlated with rapid fluctuations in power and/or head pressures. Many times surging occurs when steam, formed during the high temperature extrusion, escapes down the screw toward the feed hopper. The flow of the steam disrupts the compacted food in the channel and momentarily reduces the discharge from the extruder. Such a condition is sometimes called "steaming back" and is recognized by the presence of steam at the feed hopper. Surging can sometimes be stopped by cooling the jackets at the feed section, temporarily cooling the discharge of the extruder and/or increasing feed rate. Both of these latter actions often reduce the extruder temperature, allow the screw to properly refill with food and return to normal operating conditions.

Sometimes rapid changes in extrusion processing conditions occur that require immediate and drastic action if machine damage or difficult cleanup situations are to be avoided. The loss of a feed ingredient causes severe upsets. A drop in water or steam input rate can be the most serious because the feed becomes very dry, causing the extruder to plug and/or the motor overload. In these cases, the operator must quickly divert the dry feed and restore the addition of moisture. Overcompensation with excesses of water is desirable rather than allowing the extruder to become plugged.

E. Shutdown

In many senses, the shutdown procedures for extruders are the reverse of those for start-up. Steam to the jackets, preconditioner, and/or barrel is shut off. Excess moisture is added to the feed until the discharge temperature is reduced below 100°C. At this point, the feed is shut off but the extruder is allowed to run until wet, cool product ceases to emerge from the die.

Sometimes the screw is then shut off and the die plate removed. Care needs to be exercised in removing the die plate because, if the extruder is hot, pressures can be released violently when the head bolts are loosened, causing a potential hazard to the operators. If pressure exists behind the die while it is being removed, then a chain or other type of restraint needs to be placed on the head to keep it in place during disassembly until the pressure is relieved. After die removal, the screw can be started and the remaining material in the channel can be augered out. To facilitate cleanup, dry grits or whole soybeans are sometimes fed with the open discharge to aid in pushing out any remaining material and cleaning the screw channel.

After all drives and flows of water, steam, and feed have been shut off, the extruder is commonly disassembled for cleaning and inspection. If the barrel is segmented, it is disassembled and the screw sections dismantled. Some barrels are split so that they hinge open quickly exposing the screw. If the screw is in a single piece, then it is extracted for washing and the barrel inspected.

It is an all too common practice to use excessive force or prying techniques when disassembling extruder parts that may have become stuck. The result can be physical damage to the parts which seriously affect subsequent operations. Metal hammers, crow bars, or other hardened tools should not be used to pound or pry the screw or barrel, if damage is to be avoided. Wooden blocks should be placed between the hammer or pry before force is exerted. If a screw is stuck, the use of hydraulic pullers or screw jacks allow the application of even sustained forces, rather than the abrupt forces accompanying sharp blows which cause the most serious damage.

Providing special containers for bolts or other small parts of the extruder lessens the chance of their loss. Also, it reduces the possibility for a bolt or other metal part to accidentally fall into the extruder, causing severe damage when the machine is started.

Many times the extrusion screw, barrel sections, and dies are soaked in a vat as part of the cleaning process. High pressure washing devices are available to clean intricate dies and have found wide acceptance in the macaroni industry. Hard metal objects or drills should be avoided in cleaning dies and other delicate parts where changes in surface characteristics or scratches can lead to altered operations. Once parts are dried, they are sometimes placed in vegetable oil for storage or dried.

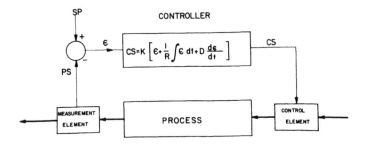

FIGURE 2. Three mode process controller operating on a simple feedback control loop.

III. CONTROL

Chapter 6 focused on extrusion variables and their measurement and in certain cases their control. This section on control will go beyond those discussions and focus on control methodology and techniques which have been found useful on food extruders.

A. Types of Control

Three types of control procedures are available for food extruders as defined below.

1. Manual Control

Many feeders are of the open loop or inferential type. Manual adjustment of the speed of a screw to control dry ingredient rate or opening of a valve to adjust a liquid rate are examples of open loop control. Often extruders operate with an operator making adjustments to his control elements based primarily on observations about the operation of the extruder or intermittent checks on feed rates by catching a timed sample. If they are not carefully performed (which requires the collection of a large sample), feed checks can show more variation in flow rates than actually exist.

The difficulty with open loop control systems is that they tend to drift with time and only by using regular calibration can any uniformity of operation be maintained. Many managers argue that open loop systems are satisfactory and sufficient because an operator has to be present anyway. The natural variability of food ingredients are such that extruder fluctuations requiring operator attention should be expected, and it is argued that this is beyond the capability of many control systems. Such operating philosophies tend to be self-fulfilling because the manual or open loop control of extruders often leads to extruder fluctuations.

2. Feedback Control

Feedback or closed loop control is practiced when changes in the control element are made on the basis of an actual measurement of the controlled variable to achieve the desired condition. The feedback control system first compares the process signal (PS) with the set point (SP) or desired process condition. The difference in these two signals is termed the error (ϵ). The error signal is manipulated by the controller to determine the control signal (CS) which adjusts a control element and moves the process toward the desired set point. Such a feedback control system is shown in Figure 2.

Feedback control systems commonly come with various modes of control. Typically, there are as many as three modes of control termed proportional (gain), reset (integral), and derivative (rate). In the time domain, the controller output is given by

$$CS = K\left[\epsilon + \frac{1}{R}\int\epsilon\,dt + D\frac{d\epsilon}{dt}\right] \qquad (8.1)$$

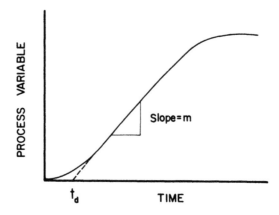

FIGURE 3.   Response curve for controlled process where
manipulated variable has a step change ΔMV.

where CS = control signal, K = proportional gain, ε = error signal, R = reset time, min, t = time, min, and D = derivative time, min. On a three mode controller, K, R, and D are all adjustable. When K is increased, the CS will proportionally increase. The reset or integral action will cause the CS to change as long as ε exists, so that the controller will move toward the SP and eliminate any permanent offset between the SP and PS. R, the reset time, is the time required for the integral portion of the controller output to become equal to the proportional part at a constant error.

Derivative response is a function of the rate of change in the error signal with time, dε/dt. D, derivative time, is the time required for the controller output to become equal to twice the proportional response.

To summarize, proportional gain generally specifies the response of the controller with higher gains giving larger responses but greater potential for process oscillation. Integral control eliminates steady state offset between the process operating point and the set point. Derivative mode allows higher proportional gains and the system to respond to anticipated changes in the process operating point.

The optimal settings of the control constants depends on the response characteristics of the extrusion system with time. When no model of the time response of the system is known, some approximate techniques have been worked out for determining the "best" control settings. First a step change in the manipulated variable (MV) is imposed on the system and the change in PS with time is measured as shown in Figure 3. From the maximum slope, m, the rate of change per unit-step change of the MV is calculated as

$$R_m = \frac{m}{\Delta MV} \tag{8.2}$$

where R_m = rate of change per unit — step change, m = maximum slope of response curve, and ΔMV = change in manipulated variable. The best controller settings based on the reaction-curve process tests are given in Table 1 as a function of R_m and t_d.

3. Feedforward Control

Feedforward control differs from feedback control in that process changes at the input are measured and corrective action is taken before the controlled variable actually responds to the input variable disturbances. The application of feedforward control requires a thorough knowledge of the response of the extruder system to changes

Table 1
CONTROLLER SETTINGS BASED ON RESPONSE
CURVE TO STEP CHANGE IN PROCESS

Control action	Proportional grain (K)	Reset time, R (min)	Derivative time, D (min)
Proportional	$1/R_m t_d$	—	—
Proportional + integral	$0.9/R_m t_d$	$3 t_d$	—
Proportional + integral + derivative	$1.2/R_m t_d$ to $2.0/R_m t_d$	$2.5 t_d$ to $2.0 t_d$	$0.5 t_d$ to $0.3 t_d$

Note: $R_m = m/\Delta MV$, t_d = dead time.

in input process variables so that corrective measures will be appropriate. Feedforward control is very suitable for systems with long dead times which make feedback control relatively insensitive and the controlled response very slow.

B. Types of Controllers
Two types of feedback controllers are available and widely used as described below.

1. Analog Controllers
Analog controllers work on a continuous analog signal from the process, perform the control calculations mechanically or electronically on the error signal, and output an analog control signal. Both pneumatic and electronic analog controllers are available. In the case of pneumatic systems, all analog signals have been standardized so that the full scale range (0 to 100%) is 0.02 to 0.10 MPa (3 to 15 psi). It is convenient to use pneumatic controllers for flow control loops because the control element is typically a valve which has a pneumatic actuator. Input signals for these loops are either temperature or flow, measured by an orifice, which is conveniently converted to pneumatic signals.

Electronic controllers have been standardized so they receive and output standard 4 to 20 ma analog signals for a full range of 0 to 100%. The control calculations on the error signal are performed with an operational amplifier network.

Both types of analog controllers are rugged and dependable but require scheduled periodic maintenance. The major maintenance problems involve the sensors and control elements. Plant maintenance personnel can be trained to perform these functions.

2. Digital Control
The application of digital computers to process control has increased rapidly over the past 15 years. Digital control becomes economic when the numbers of control loops increases and/or the control requirements become more complex than can be performed by simple analog controllers. Feedforward control is a good example of a control methodology which may be difficult or expensive to perform with specialized instrumentation, but is relatively straight forward with a digital system.

With direct digital control, the digital computer performs all control calculations for individual loops on an intermittent basis. A schematic of a direct digital control system is shown in Figure 4. Most process signals are analog and so must be converted to their digital equivalent with an A/D (analog to digital) converter. Upon command from the computer, the input signals are scanned and digitized. Calculation of the correct output control signal is done numerically with a program on the timed com-

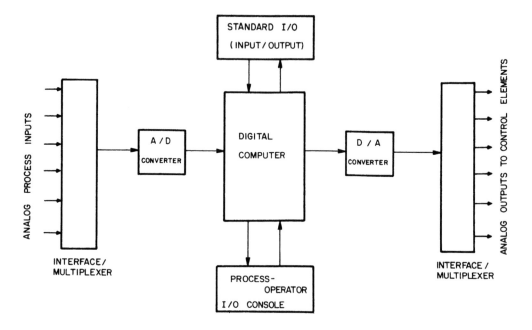

FIGURE 4.   Schematic of a direct digital process control system.

mand of the digital computer. The frequency of scanning of input and output control signals depends upon the speed of response of the process and is programmable. The digital output signal frequently goes to an analog control element so must be converted from digital to analog with a D/A converter. The interface or multiplexer involved has computer-controlled switches which isolate the process from the computer and hold the input and output for the computer while it is performing other functions.

The multipurpose digital computer has come down dramatically in cost and increased its speed so that digital control is currently very cost competitive. In addition to control, the computer can also easily log process data, generate process reports, alarm operators, etc. with very little additional expense.

C. Individual Control Loops

Figure 5 shows the use of three individual control loops to control the dry ingredient, water, and steam feed rates to and extruder. Such a system should assure the constant input rates of all the important feeds. The set points for the individual controllers are adjusted by the operator in response to his knowledge of the reaction of the extruder to changes in these independent variables. To assist the operator in accessing overall extruder operations, indicators are available for power, product discharge temperature, and pressure behind the die.

Clearly, it can be seen that the operator must be present to make necessary set point changes as required by variations in ingredients, ambient conditions, or extruder wear. The advantage of individual feedback controllers on each feed ingredient is that the desired rates are maintained within rather narrow limits, which is much more difficult to assure with open loop or inferential control devices.

D. Temperature Control Loop

A common method to control the discharge extrudate temperature with a jacketed extruder is shown in Figure 6. Here, the feed and water rates are maintained with individual feed control devices. The temperature of the extrudate is measured just be-

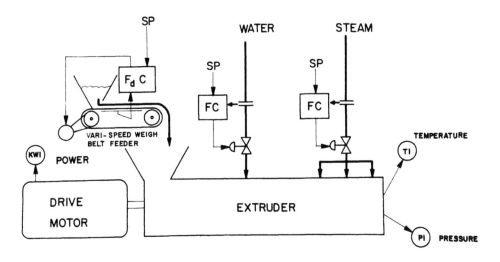

FIGURE 5.   Individual control loops for feed, water, and steam rates.

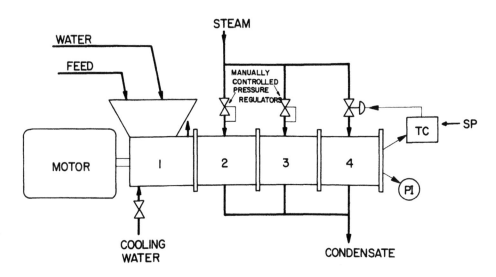

FIGURE 6.   Temperature controller (TC) adjusting steam pressure in fourth barrel jacket to maintain extrudate temperature. Steam pressure in jackets two and three is manually controlled.

hind the die plate and a temperature controller adjusts a steam valve and, correspondingly the steam pressure in the last or fourth jacket on the barrel. In this manner, the final jacket temperature can be made to compensate for minor fluctuations in the final extrudate temperature by heat transfer between the steam and extrudate.

On the middle two jackets of the system shown, the steam pressure is controlled manually with simple steam pressure regulators. It is common to have $P_2 < P_3$ so that the temperature of the jackets increases progressively down the barrel. The precise pressure in these jackets is adjusted on the basis of experience. Water is usually circulated in the first jacket to improve ingredient feeding.

The temperature control system shown can serve to maintain a constant extrudate temperature when the temperature fluctuations are rather modest. The following example quickly shows that the amount of heat transferred between a jacketed barrel and extrudate is quite small compared to the total heat requirements for large extruders.

E. Example 1

Heat is being added to a food material in an extruder from steam condensing at 4 bar absolute gauge in a jacket surrounding the barrel. The barrel is 15.25 cm I.D. and made from 420 stainless steel (SS) having a thickness of 19.0 mm. The surface heat transfer coefficient, h, for the food dough is 280 $W/m^2{}^\circ K$ and for the condensing steam is 2200 $W/m^2{}^\circ K$. As the food passes through a 0.91m section of the screw, calculate the quantity of heat transferred if the average dough temperature is 105°C. What is the expected rise in temperature of the extrudate due to the heat transferred if the heat capacity, c_p, is 2.5 kJ/kg°K and flow rate, $\dot{m}$, is 850 kg/hr.

For the solution:

Base U, the overall heat transfer coefficient on the inside area, or

$$U_i = \frac{1}{\dfrac{1}{h_i} + \dfrac{xD_i}{kD_{\ell m}} + \dfrac{D_i}{h_o D_o}}$$

$$k = 24.9 \ W/m^\circ K$$

$$D_{\ell m} = \frac{D_o - D_i}{\ln \dfrac{D_o}{D_i}} = \frac{19.05 - 15.25}{\ln \dfrac{19.05}{15.25}} = 17.08 \ cm$$

$$U_i = \frac{1}{\dfrac{1}{280} + \dfrac{0.019(15.25)}{24.9(17.08)} + \dfrac{15.25}{2200(19.05)}} = 217 \ W/m^2{}^\circ K$$

Temperature of saturated steam at 4 bar = 144°C.

Calculate heat transferred,

$$q = U_i A_i \Delta T = (217)(\pi)(0.1525)(0.91)(144 - 105) = 3690 \ W = 13.3 MJ/hr$$

Calculate temperature rise in extrudate from heat transferred from last jacket.

$$\Delta T = \frac{q}{\dot{m}c_p} = \frac{13.3(1000)}{850(2.5)} = 6.25°C$$

This example clearly shows that jackets on large extruders are relatively ineffective as a means of adding large quantities of heat to the extrudate. To increase heat input through heat transfer, heating the extrusion screw has been found an effective way in addition to heating the barrel jackets.

F. Cascade Control System

More elaborate control systems have been used on food extruders to reduce operator intervention and improve product uniformity. One such system is shown in Figure 7. Here, the primary control variable is the motor power. In an extrusion system where there is direct injection of water and steam, motor power is a complex function of feed rate and the relative proportions of water and steam.

The system used is commonly called a cascade control system. The dry feed ingredient rate is set as a function of the motor power. A signal proportional to the kWs drawn by the motor is fed to a ratio controller, whose output is the set point for the

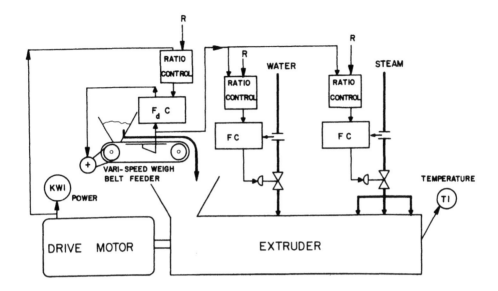

FIGURE 7. Cascade control system where feed rate changes are proportional to power and water and steam are proportional to the feed rate.

feeder. In this manner, the feed rate is adjusted automatically to maintain the desired maximum power from the motor, thus maximizing production. As ingredient or other changes occur, feed rate adjustments maintain the load on the extruder motor.

Similarly, to maintain the water and steam rates in direct proportion to the adjusted feed rates, the cascade control system is again used on the water and steam control loops. The actual feed rate signal is fed through two separate ratio controllers to generate the set points for the water and steam flow controllers, always maintaining the preset ratio between these feed streams to the extruder. The operator can alter the ratio between the elements in the control scheme by adjusting the inputs, R, to the various ratio controllers.

IV. MAINTENANCE

The maintenance of extrusion equipment varies with the specific type of extruder and the types of products being extruded. In general, maintenance includes proper lubrication, bearing inspection, replacement of worn screws and barrel liners, calibration of instrumentation, and sharpening and adjustment of cutters. Preventative maintenance schedules need to be developed for extrusion equipment and a responsible person needs to keep records showing when maintenance was performed, parts replaced, or adjustments made. Low maintenance costs are about $0.30/MT while high maintenance costs would be $1.50/MT.

The scope of the maintenance of extruders is discussed below.

A. Lubrication and Inspection

The regular lubrication of bearings and gear reducers in the drive train of the extruder and the bearings in the cutter assembly is essential for long life of these components. All extruder manufacturers give lubrication instructions and specify the types of lubricants which are required. Either the extruder operator or (more commonly) a special person from the plant mechanical staff is responsible for the scheduled maintenance.

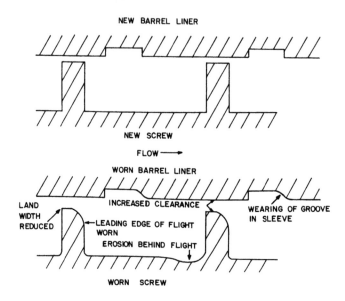

FIGURE 8.   Typical wear patterns shown on cross-section of extrusion screws and barrel liners.

During lubrication, it is normal practice for all bearings to be checked for wear and all grease and oil seals inspected for leaks. Any detected problems should be repaired before the extruder is started. Cleanliness during lubrication is essential to prevent dirt, food, or water from contaminating the lubricant and nullifying its effectiveness.

The life of the main bearings and thrust bearings on the extrusion screw is usually a minimum of 5000 hr but may exceed 10,000 hr. Seals on these bearings must be kept in perfect condition so that grease or oil does not have a chance to enter the food material or conversely, food does not contaminate the lubricant and ruin the bearings.

B. Screw and Barrel Wear

The wearing of extrusion screws and barrel liners occurs with time, requiring that these components be replaced. Since large quantities of mechanical energy are dissipated in the extrusion screw, particularly in the region of the lands on the screw or grooves on the barrel, wear of these components is unavoidable. Keeping ingredients free from dirt or foreign abrasive matter and minimization of running the extruder when it is empty will reduce the rate of wear and increase the life of the extruder components to a maximum.

Several types of wear occur on the extruder components. Figure 8 shows how the flights on a screw and the grooves on the barrel liners become worn with time. Worn components reduce extrusion rate and the ability to achieve required extrusion temperatures and may cause surging and nonuniformity of the extrudate. Careful measurements on all new or replacement screws or barrel liners before installation and after operation will provide data on when replacement of worn parts must be made to maintain satisfactory extrusion operations.

Rather than replacing worn parts with new parts, it is normal practice to rebuild worn parts. Welding and grinding techniques are frequently employed for this purpose. The worn screw is cleaned and the worn surfaces built up with stellite or other hard surface welding rods. Once the dimensions exceed those of new components, grinding of the surfaces to original specifications completes the rebuilding process. After grinding, further hardening processes may be used to bring the surface hardness up to 50 to 65 Rockwell C. A smooth polished channel on the rebuilt screw is necessary to achieve maximum extrusion output of the rebuilt screw.

Many screws can be successfully rebuilt three to five times. Each rebuilding typically costs about one-fourth the price of a new screw. Extruder manufacturers and other firms specialize in the component rebuilding process.

Barrel liners are also built up and reground to renew worn surfaces. Prior to the rebuilding, however, sharpening the leading edge of the groove will greatly improve the performance of the liner or sleeve. Small grinding wheels and lathes can sharpen the leading edge by slightly expanding the size of the groove. Such sharpening can be done three to five times before rebuilding is required but each time it is done, extrusion throughput usually decreases because of the increased groove area causing greater pressure flows. Over time, the barrel liner also becomes worn at its surface increasing clearance and requires replacement.

The life of screws and barrel liners varies extensively with extrusion conditions. With some collet extruders, barrel liners may last only 50 hr with screws lasting three to four times as long. On pet foods containing abrasive meat and bone meal, extrusion screws and barrels typiclly have to be replaced between 500 and 1000 hr. The extrusion of high moisture foods with little viscous dissipation of mechanical energy causes minimum wear and components may last up to 6000 hr.

C. Instrument and Controller Maintenance

At least once a month, all sensors, instruments, and controllers need to be thoroughly checked, cleaned, calibrated, and adjusted. When failure of one of these components is noticed, immediate maintenance is required. The long term drift or loss of calibration of instruments can often go undetected leading to operational anomalies and the loss of confidence in the extruder controls.

Many food factories hire an outside specialist to maintain, calibrate, and repair the instrumentation and controls. Such a practice is practical because the frequency of preventative maintenance is low and the level of specialized skill required is high. Blanket service contracts can often be written with instrument suppliers or other specialized firms.

D. Dies and Cutters

Die inserts wear over time and require replacement. Often the evidence of wear is slow and goes undetected until the packaging department begins to complain of packaging underfill due to thicker, heavier pieces being extruded. The use of a micrometer to regularly measure the critical dimensions of extruder pieces is one way to maintain uniformity and signal the time for die or insert replacement. Obviously, misshapen pieces or other visable defects in the extruded products are also indications of die wear.

Cutter blades must be exchanged with freshly sharpened blades relatively often. When precision cutting is required, blade changes every two hours may be necessary. In other cases blades may operate for days.

The sharpening of knife blades requires a great deal of care and precision. The bevel on the knife and angle of the sharpened portion are all critical. To accomplish precision sharpening, it is common to use a special jig with a well dressed grinding wheel.

The bearings on the rotating cutter assembly must be of precision quality and in good operating order. Any play in the bearings will result in varying clearances between the blades and die face causing rapid blade wear and/or poor cutting. It is necessary to frequently test the cutter bearings to determine their operating capability.

E. Spare Parts

Critical spare parts must be stocked and available for repairs or replacements as

Table 2
PRODUCTION COST SHEET FOR AN EXTRUSION PROCESS. BASIS _____ DAYS OPERATION

1.0 Manufacturing cost
    1.1 Direct production cost
        1.1.1 Raw materials (total raw materials requirements including losses for dehulling, moisture loss, product shrinkage, etc.)
        1.1.2 Credit for sales of offal and scraps
        1.1.3 Operating labor (salaries and benefits)
        1.1.4 Direct supervisory labor (salaries and benefits)
        1.1.5 Packaging labor (salaries and benefits)
        1.1.6 Packaging materials and supplies
        1.1.7 Utilities (gas, steam, water, electricity)
        1.1.8 Maintenance and repairs (salaries and benefits)
        1.1.9 Operating supplies (spare parts and miscellaneous supplies)
        1.1.10 Laboratory charges
        1.1.11 Patents and royalties
    1.2 Fixed charges
        1.2.1 Depreciation — equipment and facilities
        1.2.2 Interest on borrowed capital
        1.2.3 Local taxes
        1.2.4 Insurance
        1.2.5 Rent, etc.
    1.3 Overhead costs — general plant upkeep, payroll overhead, medical services, safety, lunch room, recreation, etc.
2.0 General expense
    2.1 Administration costs — executive salaries, clerical wages, legal fees, office supplies, communications
    2.2 Distribution and selling costs — sales office, salesmen, shipping, advertising
    2.3 Research and development
3.0 Total production cost — manufacturing cost of general expense

needed. These parts include new bearing assemblies, barrel liners, screws, die inserts, and cutter head and blades. Since many of the wearing parts of an extruder can be rebuilt, sufficient quantities are needed to assure the availability of spares while some parts are out for rebuilding. In the case of screws or barrel liners, this usually means a minimum of two spares. If wear is rapid, larger numbers of spares will be required.

V. OPERATING COSTS

Accurate processing costs are necessary for management control and product costing. Because the extruder is often part of a large plant making several products, the precise assignment of costs to an individual line or group of extruders is difficult. To cover some of these difficulties, careful monitoring of the extrusion line is done for a week or so to develop good estimates of utilities, ingredients, labor, packaging, maintenance, loss, etc. as a function of production. Standard cost accounting methods are used to assign and amortize fixed charges, overhead costs and general expense. These short term measurements are helpful in establishing true production costs but provide little input for management control of on-going costs.

A typical total cost analysis sheet for an extrusion process is given in Table 2. The direct product cost is of most interest to the plant manager. Careful material balances are required to accurately know raw material costs. Data on raw material usage and product produced should be kept continuously and checked against a standard. Quality control needs to run standard tests to see that all ingredients meet specification in terms of quality and moisture to control costs. All food processing plants produce some offal or off-specification product which can normally be sold and salvage credit obtained.

For a typical extruded food product cost analysis, raw materials (ingredients) are about 35 to 60% of the cost, labor 5 to 10%, all packaging costs 25 to 50%, utilities 5 to 10% and all other costs including maintenance 5%. The breakdown between raw materials and packaging costs varies considerably depending on the type of product and the size of the finished product package. Very small packages increase the relative packaging costs tremendously.

For high value extruded items such as cereals, snacks, and some pet foods, the total production cost may be 50 to 70% of the retail sales price. The difference is wholesaler, retailer, and manufacturer profit. On higher volume commodity items, total production cost would be closer to 70 to 80% of retail sales price.

VI. LEAST ENERGY COST OPERATION

Energy costs are rapidly increasing and becoming a larger portion of the production costs. The energy costs and requirements for extrusion can be varied significantly by altering the extruder operation. Within some limits, it is possible to vary the source of the input energy to an extrusion system by altering the moisture content of the extrudate. Dry extrusions tend to have a large fraction of their energy input from mechanical dissipation, while wet extrusions have a larger portion of their input energy requirements from direct steam injection or heat transfer through the jackets. Because the cost of heat varies with the energy source, it is possible to minimize the energy costs of extrusion by adjusting the extrusion operation knowing the operating characteristics of the system and the relative energy costs. Typically, electrical energy used for the drive is much more expensive, on a heat equivalent basis, than energy from steam. It would appear that higher moisture extrusion, with most of the energy inputs coming from condensing steam, would be the least expensive.

The actual energy utilization and cost analysis as a function of extrudate moisture content is much more complex because at higher moisture contents, increased product drying energy needs are incurred to remove the added water and bring the product to a safe storage moisture. Drying is usually done with natural gas or propane; consumption and cost must also be calculated as a function of extrudate moisture content to determine minimum energy cost.

Since energy costs vary from one region of the country to another and the effect of moisture on the extrusion operation cannot be generalized, it is impossible to generalize about optimal moisture levels for food extrusion. The following example does illustrate the trade-offs discussed above in a practical problem directed toward calculating the optimal moisture level for any extrusion process.

A. Example 2

An extrusion system is to operate at the finished product rate of 2000 kg/hr of pet food at a moisture of 10% WB. The extrusion system used is pictured schematically below.

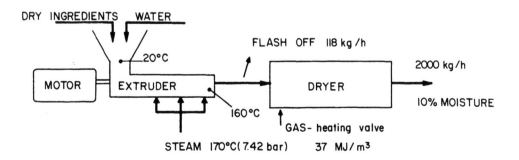

The extruder can be operated dry with very little moisture addition where most of the energy required to heat the product comes from viscous dissipation of the input mechanical shaft energy. Alternatively, the process can be operated wet where most of the energy is added through steam injection to the barrel of the extruder and incorporated directly into the product. Pilot plant data have been gathered (Figure 9) which show the percentage of total energy to heat the product coming from viscous dissipation of mechanical input energy as a function of moisture content within the extruder. Although for wet extrusion, mechanical input energy is less, drying requirements are increased to bring the final product moisture to the 10% specification.

For the purpose of calculation, the specific heat of the pet food can be considered a simple function of the moisture content of extrusion.

$$c_p = 2.72\, x_m + 1.55,\ kJ/kg°K$$

where x_m = fraction moisture. The endothermic heat of gelatinization of the starchy component of the ingredients is 6.6 kJ/kg of dry ingredients. Also, assume moisture loss at the die equals 5% due to flash off of steam as product emerges. To remove 1 kg water in the dryer requires the addition of 3.5 MJ of energy.

Calculate the optimal moisture content for the extrusion process to achieve minimum energy cost if electricity is $0.03/kWhr, steam $3.30/Mg, and gas $4.40/hm³.

For the solution, the calculation procedure is outlined below.

2000 kg/hr finished product at x_p = 0.10. The heating requirement is the sensible heat rise in the product and endothermic heat of reaction.

Results are summarized in Table 3 at the end of the problem solution.

1. Calculate water in product for varying extrusion moistures (x_m) which will yield P kg/hr finished product at moisture x_p.

$$P(1 - x_p) = kg\ dry\ ingredient/hr = 1800\ kg/hr$$

$$W = kg\ water/hr$$

$$\frac{W}{P(1 - x_p) + W} = x_m$$

$$W = \frac{P(1 - x_p)(x_m)}{1 - x_m}$$

2. Calculate total flow, $\dot{m}$, through extruder at x_m.

$$\dot{m} = 1800 + W$$

3. Calculate total heat requirement to heat extrusion mixture to a discharge temperature of 106°C.

$$q_T = \dot{m}c_p\Delta T + P(1 - x_p)\Delta H° = \dot{m}(2.72x_m + 1.55)(160°C - 20°C) + 1800(6.6)$$

4. Estimate total energy input coming from the viscous dissipation of mechanical energy for each x_m (see Figure 9).
5. Calculate mechanical energy input/hr.

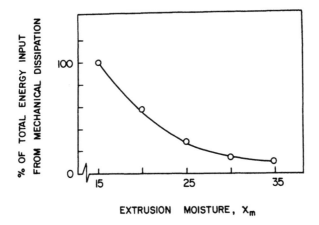

FIGURE 9. Data showing the percentage of total energy to heat product coming from viscous dissipation of mechanical input energy as a function of moisture content within the extruder.

Table 3
RESULTS FROM CALCULATIONS

x_m	1 W (kg/hr)	2 $\dot{m}$ (kg/hr)	3 $q_r \times 10^{-3}$ (kJ/hr)	4 % mech	5 $q_m \times 10^{-3}$ (kJ/hr)	6 $q_s \times 10^{-3}$ (kJ/hr)
0.15	318	2118	593	100	593	0
0.20	450	2250	672	57	383	289
0.25	600	2400	761	28	213	548
0.275	683	2483	811	21	170	641
0.30	771	2571	864	14	121	743
0.35	969	2769	982	10	98	884

x_m	7 $\dot{m}_s$ (kg/hr)	8 $\dot{m}_w$ (kg/hr)	9 $\dot{m}_D$ (kg/hr)	10 Mech ($/hr)	10 Steam ($/hr)	10 Gas ($/hr)	Total ($/hr)
0.15	0	0	0	4.95	0.00	0.00	4.95
0.20	138	−6	132	3.19	0.46	0.55	4.20
0.25	262	20	282	1.78	0.86	1.17	3.81
0.275	306	59	365	1.42	1.01	1.52	3.95
0.30	355	98	453	1.01	1.17	1.89	4.07
0.35	422	229	651	0.82	1.39	2.71	4.92

$$q_m = q_T \left(\frac{\% \text{ mech.}}{100} \right)$$

6. Calculate steam energy input/hr.

$$q_s = q_T - q_m$$

7. Calculate steam requirement, $\dot{m}_s$, kg/hr.

$$\dot{m}_s \;=\; q_s/\Delta h_{fg}$$

$$h_g \;=\; 2769 \text{ J/g} \;, \text{steam at } 170°C \;(7.92 \text{ bar})$$

$$h_f \;=\; \underline{675 \text{ J/g}} \;, \text{water at } 160°C$$

$$\Delta h_{fg} \;=\; 2093 \text{ J/g}$$

8. Calculate water addition, $\dot{m}$, kg/hr.

$$\dot{m}_w \;=\; \dot{m} \;-\; (2118 + \dot{m}_s)$$

where 2118 kg/hr is $\dot{m}$ for raw ingredients at $x_m = 0.15$.

9. Calculate water removed during drying.

$$\dot{m}_D \;=\; \dot{m}_s \;+\; m_w$$

which is the extra steam and water added for extrusion at $x_m > 0.15$.

10. Cost of energy, $/hr.

   a. Mechanical

$$\frac{q_m, \text{kJ}}{\text{hr}} \left| \frac{1 \text{ MJ}}{10^3 \text{kJ}} \right| \frac{1 \text{ kWhr}}{3.6 \text{ MJ}} \left| \frac{C_E, \$}{\text{kWhr}} \right. \;=\; 0.000278 \, q_m C_E \;=\; \$/hr$$

   b. Steam

$$\frac{\dot{m}_s, \text{kg}}{\text{hr}} \left| \frac{1 \text{ Mg}}{10^3 \text{kg}} \right| \frac{C_s, \$}{\text{Mg}} \;=\; (0.001) \, (\dot{m}_s) C_s \;=\; \$/hr$$

   c. Gas

$$\frac{\dot{m}_D, \text{kg}}{\text{hr}} \left| \frac{3.5 \text{ MJ}}{\text{kg}} \right| \frac{\text{m}^3}{37 \text{ MJ}} \left| \frac{\text{hm}^3}{10^2 \text{m}^3} \right| \frac{C_G, \$}{\text{hm}^3} \;=\; 0.000946 \, \dot{m}_D C_G \;=\; \$/hr$$

Plotting the hourly energy cost as a function of x_m gives Figure 10. The optimal extrusion moisture is $\sim$25% and the minimum energy costs are $3.81/hr.

A more precise solution could be obtained by writing an equation for total energy cost as a function of x_m. This equation could be differentiated (dC/dx_m) and the differential set to 0 to determine the optimal moisture for extrusion. In the solution shown, this approach was not used because all functional relationships were not known (i.e., Figure 9) and the resulting differentiated equation could not be solved easily when set equal to 0.

To thoroughly examine the conditions in Example 2 and calculate the true least-cost moisture, the costs of all the equipment would have to be also included. For example, the increase in direct steam injection would require a larger boiler and related equipment. The higher moisture product would also require a larger dryer. Both of these items increase the capital cost requirements of the plant and their costs, including the value of the money, would have to be amortized as an increased operating cost for the extrusion process. Conversely, the capacity of extrusion equipment increases and drive motor size decreases as the moisture content of the extrudate increases. Both of

185

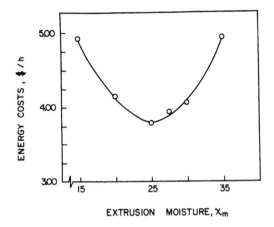

FIGURE 10.   Energy costs vs. extrusion moisture.

these factors would reduce the cost of the extruder and a cost credit could be given to the high moisture extrusion operations.

The incorporation of these amortized capital costs are beyond the scope of Example 2 but the reader must understand that total hourly operating costs would have to include energy as well as capital costs to determine the least processing cost. For smaller capacity systems, cost analysis shows dry extrusion has lower production cost because it eliminates the relatively high cost auxillary equipment. For larger capacity systems, high moisture extrusion, using a significant fraction of the total energy input from condensing steam, appears optimal.

APPENDIX 1

SYMBOLS

A	area		g_z	gravity force in z direction
a	$-(Q_p/Q_d)$, ratio pressure flow to drag flow		H	distance between screw root and barrel
			H_1	height of channel at entrance, Equation 4.39
a	acceleration		H_2	height of channel at discharge, Equation 4.39
B	axial channel width			
B	constant		H_g	height of barrel groove
b	constant		H_c	channel height
b	axial flight width		h	convective heat transfer coefficient
C	half of slit thickness		h_i	inside convective heat transfer coefficient
C	concentration			
C.R.	compression ratio		h_o	outside convective heat transfer coefficient
C.S.	control signal			
c	half thickness of slit element		j	number of grooves in parallel
c_p	specific heat		K	constant
D	barrel diameter		K	ratio of forces on solid bed, Equation 4.72
D_r	root of screw diameter			
D_s	screw diameter		K'	constant
E	energy input		K"	constant
E(t)	residence time distribution function		k	thermal conductivity
E_H	viscous energy dissipated in channel		k	reaction rate
E_k	energy to increase kinetic energy		k_r	reference reaction rate
E_p	energy to raise pressure of fluid		k_∞	apparent kinetic factor at infinite t
E_t	total net energy added to extruder		L	length of screw channel in axial direction
E_d	viscous energy dissipated in flight clearance			
			L*	equivalent length of capillary
e	flight width		L_d	length of die hole or length of probe
F	force			
F(t)	cumulative residence time distribution		l	lead
F_d	drag flow shape factor, Figure 2, Chapter 4.		M	moisture content, fraction
			M	constant, last two terms of Equation 4.71
F_{dc}	curvature correction factor for drag flow, Figure 8, Chapter 4.		m	mass
			m	constant
F_{de}	end correction factor for drag flow, Figure 6, Chapter 4.		m	consistency index
			$\dot{m}$	mass flow rate
F_{dt}	$F_d F_{de} F_{dc}$		N	speed of rotation
F_p	pressure flow shape factor, Figure 2, Chapter 4.		n	flow behavior index
			P	pressure
F_{pc}	curvature correction factor for pressure flow, Figure 9, Chapter 4.		P_1	pressure at beginning of metering section
F_{pe}	end correction factor for pressure flow, Figure 7, Chapter 4.		P_2	pressure at discharge of metering section
F_{pt}	$F_p F_{pe} F_{pc}$		P.S.	process signal
$F_{\mu d}$	drag flow viscosity factor, Figure 15, Chapter 4.		p	number of channels in parallel
			Q	volumetric flow rate
$F_{\mu p}$	pressure flow viscosity factor, Figure 15, Chapter 4.		Q_1	volumetric flow per unit slit width
f	function of		Q_d	drag flow
f	constant, Equation 4.31		Q_{do}	flow through each die hole
f_b	coefficient of friction — food and barrel		Q_g	leakage flow through barrel groove
			Q_p	pressure flow
f_s	coefficient of friction — food and screw		Q_o	volumetric flow of solids
			q	heat flux
G	constant		R	gas constant 8.314 J/g mol °K
G_z	dimensionless pressure flow for power law fluid		R	radius of probe or die
			R_H	hydraulic radius
g_c	constant		R_i	inside radius of annular die
g_x	gravity force in x direction		R_o	outside radius of annular die
g_y	gravity force in y direction		r	radius of element

S.P.	set point	v_x	velocity in x direction
s	cross sectional area	$v_x(y)$	velocity in x direction at y
T	temperature (absolute)	$v_x(yc)$	velocity in x direction at yc
T_L	temperature at end of probe	v_y	velocity in y direction
T_b	temperature of barrel	v_z	velocity in z direction
T_r	reference temperature	$v_z(y)$	velocity in z direction at y
t	time	v_a	velocity of particle in axial direction
t(y)	residence time for particle at position y	$v_a(y)$	velocity of particle at position y in axial direction
t_f	fraction of time spent in upper portion of channel	$\bar{v}$	average velocity
t_o	minimum residence time in screw occurring at y/H = 2/3	$\bar{v}_a$	average velocity of particle in axial direction
t_o	time at beginning of an interval	$\bar{v}_a(y)$	average velocity of particle at position y in axial direction
t_u	time particle spends in upper portion of channel	W	channel width
t_l	time particle spends in lower portion of channel	W_g	width of barrel groove
$\bar{t}$	average residence time	w	slit width
U	overall heat transfer coefficient	x	cross channel direction
V	πDN, peripheral speed on tip velocity	x	independent variable
V_x	−πDNsinθ	Y	distance
V_z	πDNcosθ	y	direction perpendicular to screw root
V_o	volume of screw channel	y	dependent variable
v_{max}	maximum velocity	y	position in channel
v_r (yc)	velocity of particle at position yc in axial direction	y_c	position in lower portion of channel
		Z	channel length
		z	down channel direction

GREEK LETTERS

α	angle between groove and screw tip velocity vector	Δx	thickness
β	$(2h/kR)^{1/2}$	η	apparent viscosity
γ	dx/dy, strain	η^*	reference apparent viscosity
γ_N	normalized strain	η_0	apparent viscosity at zero shear rate
$\dot{\gamma}$	dv/dy, shear rate	η_1	apparent viscosity at high temperature
$\dot{\gamma}(y)$	shear rate at y	η_2	apparent viscosity at reference moisture
$\dot{\gamma}_H$	effective shear rate in channel, Equation 4.41	η_∞	apparent viscosity at infinite shear rate
$\dot{\gamma}_a$	apparent shear rate	η_F	apparent viscosity in units of force
$\dot{\gamma}_w$	shear rate at wall	η_m	apparent viscosity in units of mass
$\dot{\gamma}_x(y)$	shear rate in x direction at y	θ	helix angle
$\dot{\gamma}_z(y)$	shear rate in z direction at y	θ_o	angle of movement of outer surface of solid plug
$\dot{\gamma}_d$	effective shear rate in clearance, Equation 4.85	μ	Newtonian viscosity
$\dot{\gamma}$	weighted average total strain	μ_b	viscosity at barrel surface
$\dot{\gamma}^*$	normalized weighted average total strain	μ_d	viscosity of dough at die
		μ_g	viscosity of dough in groove
δ	radial clearance of screw	μ_m	$(\mu_b + \mu_s)/2$, mean viscosity
ΔE°	activation energy	μ_s	viscosity at screw root
ΔE_k	activation energy for cooking section	μ_w	viscosity of dough in clearance
ΔE_η	energy of activation for flow	ϱ	density
Δh°	heat of reaction	τ	shear stress
ΔH_{S1}	latent heat of fusion	τ_o	yield stress
ΔH_{l_s}	latent heat of vaporization	τ_s	constant in Reiner-Philippolf model
ΔP	pressure difference	τ_w	shear stress at wall
ΔT	temperature difference	τ_{yx}	shear stress perpendicular to y axis in x direction
Δt	time interval	Ψ	wetted perimeter of die
		ϕ	scale-up factor
		ω	angular speed of rotation

APPENDIX 2

CONSTANTS AND CONVERSION FACTORS

A.2.1 Force

1 dyne $= 10^{-5}$ N
1 g·cm/s^2 $= 10^{-5}$ kg·m/s^2
1 kg·m/s^2 $= 1$ N
1 lb$_f$ $= 4.4482$ N
1 g·cm/s^2 $= 2.2482 \times 10^{-6}$ lb$_f$

A.2.2 Gas Law Constant R

Numerical Value	Units
1.9872	g cal/g mol·°K
1.9872	BTU/lb mol·°R
8314.3	J/kg mol·°K
10.731	ft^3·lb$_f$/in.^{2}·lb mol·°R
1545.4	ft·lb$_f$/lb mol·°R
8314.3	m^3·Pa/kg mol·°K

A.2.3 Heat Capacity and Enthalpy

1 BTU/lb$_m$·°F $= 4.1868$ kJ/kg·°K
1 BTU/lb$_m$·°F $= 1.000$ cal/g·°C
1 BTU/lb$_m$ $= 2326.0$ J/kg
1 ft·lb$_f$/lb$_m$ $= 2.9890$ J/kg

A.2.4 Heat, Energy, Work

1 J $= 1$ N·m $= 1$ kg·m^2/s^2
1 kg·m^2/s^2 $= 10^7$ g·cm^2/s^2
1 J $= 10^7$ erg
1 BTU $= 1055.06$ J $= 1.05506$ kJ
1 BTU $= 252.16$ cal
1 kcal $= 1000$ cal $= 4.1850$ kJ
1 BTU $= 778.17$ ft·lb$_f$
1 hp·hr $= 0.7457$ kW·hr
1 hp·hr $= 2544.5$ BTU
1 ft·lb$_f$ $= 1.35582$ J
1 ft·lb$_f$/lb$_m$ $= 2.9890$ J/gk

A.2.5 Heat Flux and Heat Flow

1 BTU/hr·ft^2 $= 3.1546$ W/m^2
1 BTU/hr $= 0.29307$ W

A.2.6 Heat-Transfer Coefficient

1 BTU/hr·ft²·°F = 1.3571×10^{-4} cal/s·cm²·°C
1 BTU/hr·ft²·°F = 5.6783×10^{-4} W/cm²·°C
1 BTU/hr·ft²·°F = 5.6783 W/m²·°K

A.2.7 Length

1 in. = 2.540 cm
100 cm = 1 m

A.2.8 Mass

1 lb_m = 453.59 g = 0.45359 kg
1 lb_m = 16 oz = 7000 grains
1 kg = 1000 g = 2.2046 lb_m
1 ton (short) = 2000 lb_m
1 ton (long = 2240 lb_m
1 ton (metric) = 1000 kg

A.2.9 Power

1 hp = 0.74570 kW   1 W = 14.340 cal/min
1 hp = 550 ft·lb_f/s   1 BTU/hr = 0.29307 W
1 hp = 0.7068 BTU/s   1 J/s = 1 W

A.2.10 Pressure

1 bar   = 1×10^5 Pa (pascal) = 1×10^5 N/m²
1 psia   = 1 lb_f/in.²
1 psia   = 2.0360 in. Hg at 0°C
1 psia   = 2.311 ft H_2O at 70°F
1 psia   = 51.715 mm Hg at 0°C (ϱ_{H_g} = 13.5955 g/cm³)
1 atm   = 14.696 psia = 1.01325×10^5 N/m² = 1.01325 bars
1 atm   = 760 mm Hg at 0°C = 1.01325×10^5 Pa
1 atm   = 29.92 in. Hg at 0°C
1 atm   = 33.90 ft H_2O at 4°C

A.2.11 Standard Acceleration of Gravity

g = 9.80665 m/s²
g = 980.665 cm/s²
g = 32.174 ft/s²
g_c (gravitational conversion factor)   = 32.1740 lb_m·ft/lb_f·s²
                                          = 1 kg·m/N·s²
                                          = 1 g·cm/dyne·s²

A.2.12 Thermal Conductivity

1 BTU/hr·ft·°F = 4.1365×10^{-3} cal/s·cm·°C
1 BTU/hr·ft·°F = 1.73073 W/m·K

A.2.13 Viscosity

$1 \text{ cp} = 10^{-2} \text{ g/cm} \cdot \text{s}$
$1 \text{ cp} = 2.4191 \text{ lb}_m/\text{ft} \cdot \text{hr}$
$1 \text{ cp} = 6.7197 \times 10^{-4} \text{ lb}_m/\text{ft} \cdot \text{s}$
$1 \text{ cp} = 10^{-3} \text{ Pa} \cdot \text{s} = 10^{-3} \text{ N} \cdot \text{s/m}^2$
$1 \text{ cp} = 2.0886 \times 10^{-5} \text{ lb}_f \cdot \text{s/ft}^2$
$1 \text{ N} \cdot \text{s/m}^2 = 1 \text{ Pa} \cdot \text{s} = 1000 \text{ cp}$

A.2.14 Volume

$1 \, l = 1000 \text{ cm}^3$	$1 \text{ m}^3 = 1000 \, l$
$1 \text{ in.}^3 = 16.387 \text{ cm}^3$	$1 \text{ U.S. gal} = 4 \text{ qt}$
$1 \text{ ft}^3 = 28.317 \, l$	$1 \text{ U.S. gal} = 3.7854 \, l$
$1 \text{ ft}^3 = 0.028317 \text{ m}^3$	$1 \text{ U.S. gal} = 3785.4 \text{ cm}^3$
$1 \text{ ft}^3 = 7.481 \text{ U.S. gal}$	$1 \text{ British gal} = 1.20094 \text{ U.S. gal}$
$1 \text{ m}^3 = 264.17 \text{ U.S. gal}$	

1.2.15 Volume and Density

$1 \text{ g/cm}^3 = 62.43 \text{ lb}_m/\text{ft}^3 = 1000 \text{ kg/m}^3$
$1 \text{ g/cm}^3 = 8.345 \text{ lb}_m/\text{U.S. gal}$
$1 \text{ lb}_m/\text{ft}^3 = 16.0185 \text{ kg/m}^3$

APPENDIX 3

PHYSICAL PROPERTIES OF WATER

Table 1
DENSITY OF LIQUID WATER

Temperature		Density		Temperature		Density	
(°K)	(°C)	(g/cm³)	(kg/m³)	(°K)	(°C)	(g/cm³)	(kg/m³)
273.15	0	0.99987	999.87	323.15	50	0.98807	988.07
277.15	4	1.00000	10000.00	333.15	60	0.98324	983.24
283.15	10	0.99973	999.73	343.15	70	0.97781	977.81
293.15	20	0.99823	998.23	353.15	80	0.97183	971.83
303.15	30	0.99568	995.68	363.15	90	0.96534	965.34
313.15	40	0.99225	992.25	373.15	100	0.95838	958.38

From Perry, R. H. and Chilton, C. H., *Chemical Engineers' Handbook,* 5th ed., McGraw-Hill, New York, 1973. With permission of McGraw-Hill Book Company.

Table 2
VISCOSITY OF LIQUID WATER

Temperature		Viscosity	Temperature		Viscosity
°K	°C	(N·s/m² × 10³)	°K	°C	(N·s/m² × 10³)
273.15	0	1.7921	323.15	50	0.5494
275.15	2	1.6728	325.15	52	0.5315
277.15	4	1.5674	327.15	54	0.5146
279.15	6	1.4728	329.15	56	0.4985
281.15	8	1.3860	331.15	58	0.4832
283.15	10	1.3077	333.15	60	0.4688
285.15	12	1.2363	335.15	62	0.4550
287.15	14	1.1709	337.15	64	0.4418
289.15	16	1.1111	339.15	66	0.4293
291.15	18	1.0559	341.15	68	0.4174
293.15	20	1.0050	343.15	70	0.4061
293.35	20.2	1.0000	345.15	72	0.3952
295.15	22	0.9579	347.15	74	0.3849
297.15	24	0.9142	349.15	76	0.3750
299.15	26	0.8737	351.15	78	0.3655
301.15	28	0.8360	353.15	80	0.3565
303.15	30	0.8007	355.15	82	0.3478
305.15	32	0.7679	357.15	84	0.3395
307.15	34	0.7371	359.15	86	0.3315
309.15	36	0.7085	361.15	88	0.3239
311.15	38	0.6814	363.15	90	0.3165
313.15	40	0.6560	365.15	92	0.3095
315.15	42	0.6321	367.15	94	0.3027
317.15	44	0.6097	369.15	96	0.2962
319.15	46	0.5883	371.15	98	0.2899
321.15	48	0.5683	373.15	100	0.2838

From Bingham, *Fluidity and Plasticity,* McGraw-Hill, New York, 1922. With permission.

Table 3
HEAT CAPACITY OF LIQUID WATER (1 ATM)

Temperature		Heat capacity, c_p		Temperature		Heat capacity, c_p	
(°C)	(°K)	(cal/g °C)	(kJ/ kg·°K)	(°C)	(°K)	(cal/g °C)	(kJ/ kg·°K)
0	273.15	1.0080	4.220	60	333.15	1.0001	4.187
10	283.15	1.0019	4.195	70	343.15	1.0013	4.192
20	293.15	0.9995	4.158	80	353.15	1.0029	4.199
30	303.15	0.9987	4.181	90	363.15	1.0050	4.208
40	313.15	0.9987	4.181	100	373.15	1.0076	4.219
50	323.15	0.9992	4.183				

From Osborne, N.S. Stimsom, H. F., and Ginnings, D. C., *Bur. Stand. J. Res.*, 23, 197, 1939.

Table 4
THERMAL CONDUCTIVITY OF LIQUID WATER

Temperature			BTU/hr·ft·°F	W/m·°K
(°C)	(°F)	(°K)		
0	32	273.15	0.329	0.569
37.8	100	311.0	0.363	0.628
93.3	200	366.5	0.393	0.680
148.9	300	422.1	0.395	0.684
215.6	420	588.8	0.376	0.651
326.7	620	599.9	0.275	0.476

From Timrot, D. L. and Vargaftik, N. B., *J. Tech. Phys. (U.S.S.R.)*, 10, 1063, 1940. With permission.

195

Table 5
PROPERTIES OF SATURATED STEAM AND WATER, SI UNITS

Temperature (°C)	Vapor pressure (kPa)	Specific volume (m³/kg)		Enthalpy (kJ/kg)		Entropy (kJ/kg·K)	
		Liquid	Saturated vapor	Liquid	Saturated vapor	Liquid	Saturated vapor
0.01	0.6113	0.0010002	206.136	0.00	2501.4	0.0000	9.1562
3	0.7577	0.0010001	168.132	12.57	2506.9	0.0457	9.0773
6	0.9349	0.0010001	137.734	25.20	2512.4	0.0912	9.0003
9	1.1477	0.0010003	113.386	37.80	2517.9	0.1362	8.9253
12	1.4022	0.0010005	93.784	50.41	2523.4	0.1806	8.8524
15	1.7051	0.0010009	77.926	62.99	2528.9	0.2245	8.7814
18	2.0640	0.0010014	65.038	75.58	2534.4	0.2679	8.7123
21	2.487	0.0010020	54.514	88.14	2539.9	0.3109	8.6450
24	2.985	0.0010027	45.883	100.70	2545.4	0.3534	8.5794
27	3.567	0.0010035	38.774	113.25	2550.8	0.3954	8.5156
30	4.246	0.0010043	32.894	125.79	2556.3	0.4369	8.4533
33	5.034	0.0010053	28.011	138.33	2561.7	0.4781	8.3927
36	5.947	0.0010063	23.940	150.86	2567.1	0.5188	8.3336
40	7.384	0.0010078	19.523	167.57	2574.3	0.5725	8.2570
45	9.593	0.0010099	15.258	188.45	2583.2	0.6387	8.1648
50	12.349	0.0010121	12.032	209.33	2592.1	0.7038	8.0763
55	15.758	0.0010146	9.568	230.23	2600.9	0.7679	7.9913
60	19.940	0.0010172	7.671	251.13	2609.6	0.8312	7.9096
65	25.03	0.0010199	6.197	272.06	2618.3	0.8935	7.8310
70	31.19	0.0010228	5.042	292.98	2626.8	0.9549	7.7553
75	38.58	0.0010259	4.131	313.93	2635.3	1.0155	7.6824
80	47.39	0.0010291	3.407	334.91	2643.7	1.0753	7.6122
85	57.83	0.0010325	2.828	355.90	2651.9	1.1343	7.5445
90	70.14	0.0010360	2.361	376.92	2660.1	1.1925	7.4791
95	84.55	0.0010397	1.9819	397.96	2668.1	1.2500	7.4159
100	101.35	0.0010435	1.6729	419.04	2676.1	1.3069	7.3549
105	120.82	0.0010475	1.4194	440.15	2683.8	1.3630	7.2958
110	143.27	0.0010516	1.2102	461.30	2691.5	1.4185	7.2387
115	169.06	0.0010559	1.0366	482.48	2699.0	1.4734	7.1833
120	198.53	0.0010603	0.8919	503.71	2706.3	1.5276	7.1296
125	232.1	0.0010649	0.7706	524.99	2713.5	1.5813	7.0775
130	270.1	0.0010697	0.6685	546.31	2720.5	1.6344	7.0269
135	313.0	0.0010746	0.5822	567.69	2727.3	1.6870	6.9777
140	316.3	0.0010797	0.5089	589.13	2733.9	1.7391	6.9299
145	415.4	0.0010850	0.4463	610.63	2740.3	1.7907	6.8833
150	475.8	0.0010905	0.3928	632.20	2746.5	1.8418	6.8379
155	543.1	0.0010961	0.3468	653.84	2752.4	1.8925	6.7935
160	617.8	0.0011020	0.3071	675.55	2758.1	1.9427	6.7502
165	700.5	0.0011080	0.2727	697.34	2763.5	1.9925	6.7078
170	791.7	0.0011143	0.2428	719.21	2768.7	2.0419	6.6663
175	892.0	0.0011207	0.2168	741.17	2773.6	2.0909	6.6256
180	1002.1	0.0011274	0.19405	763.22	2778.2	2.1396	6.5857
190	1254.4	0.0011414	0.15654	807.62	2786.4	2.2359	6.5079
200	1553.8	0.0011565	0.12736	852.45	2793.2	2.3309	6.4323
225	2548	0.0011992	0.07849	966.78	2803.3	2.5639	6.2503
250	3973	0.0012512	0.05013	1085.36	2801.5	2.7927	6.0730
275	5942	0.0013168	0.03279	1210.07	2785.0	3.0208	5.8938
300	8581	0.0010436	0.02167	1344.0	2749.0	3.2534	5.7045

From Keenan, J. H., Keyes, F. G., Hill, P. G., and Moore, J. G., *Steam Tables — Metric Units*, Copyright © 1969, John Wiley & Sons. Reprinted by permission of John Wiley & Sons, Inc.

Table 6
PROPERTIES OF SUPERHEATED STEAM, SI UNITS (v, SPECIFIC VOLUME, m³/kg; H, ENTHALPY, kJ/kg; s, ENTROPY, kJ/kg·K)

Absolute pressure, kPa (saturated temperature °C)		Temperature (°C)							
		100	150	200	250	300	360	420	500
10 (45.81)	v	17.196	19.512	21.825	24.136	26.445	29.216	31.986	35.679
	H	2687.5	2783.0	2879.5	2977.3	3076.5	3197.6	3320.9	3489.1
	s	8.4479	8.6882	8.9038	9.1002	9.2813	9.4821	9.6682	9.8978
50 (81.33)	v	3.418	3.889	4.356	4.820	5.284	5.839	6.394	7.134
	H	2682.5	2780.1	2877.7	2976.0	3075.5	3196.8	3320.4	3488.7
	s	7.6947	7.9401	8.1580	8.3556	8.5373	8.7385	8.9249	9.1546
75 (91.78)	v	2.270	2.587	2.900	3.211	3.520	3.891	4.262	4.755
	H	2679.4	2778.2	2876.5	2975.2	3074.9	3196.4	3320.0	3488.4
	s	7.5009	7.7496	7.9690	8.1673	8.3493	8.5508	8.7374	8.9672
100 (99.63)	v	1.6958	1.9364	2.172	2.406	2.639	2.917	3.195	3.565
	H	2676.2	2776.4	2875.3	2974.3	3074.3	3195.9	3319.6	3488.1
	s	7.3614	7.6134	7.8343	8.0333	8.2158	8.4175	8.6042	8.8342
150 (111.37)	v		1.2853	1.4443	1.6012	1.7570	1.9432	2.129	2.376
	H		2772.6	2872.9	2972.7	3073.1	3195.0	3318.9	3487.6
	s		7.4193	7.6433	7.8438	8.0720	8.2293	8.4163	8.6466
400 (143.63)	v		0.4708	0.5342	0.5951	0.6548	0.7257	0.7960	0.8893
	H		2752.8	2860.5	2964.2	3066.8	3190.3	3315.3	3484.9
	s		6.9299	7.1706	7.3789	7.5662	7.7712	7.9598	8.1913
700 (164.97)	v			0.2999	0.3363	0.3714	0.4126	0.4533	0.5070
	H			2844.8	2953.6	3059.1	3184.7	3310.9	3481.7
	s			6.8865	7.1053	7.2979	7.5063	7.6968	7.9299
1000 (179.91)	v			0.2060	0.2327	0.2579	0.2873	0.3162	0.3541
	H			2827.9	2942.6	3051.2	3178.9	3306.5	3478.5
	s			6.6940	6.9247	7.1229	7.3349	7.5275	7.7622
1500 (198.32)	v			0.13248	0.15195	0.16966	0.18988	0.2095	0.2352
	H			2796.8	2923.2	3037.6	3.1692	3299.1	3473.1
	s			6.4546	6.7090	6.9179	7.1363	7.3323	7.5698
2000 (212.42)	v				0.11144	0.12547	0.14113	0.15616	0.17568
	H				2902.5	3023.5	3159.3	3291.6	3467.6
	s				6.5453	6.7664	6.9917	7.1915	7.4317
2500 (223.99)	v				0.08700	0.09890	0.11186	0.12414	0.13998
	H				2880.1	3008.8	3149.1	3284.0	3462.1
	s				6.4085	6.6438	6.8767	7.0803	7.3234
3000 (233.90)	v				0.07058	0.08114	0.09233	0.10279	0.11619
	H				2855.8	2993.5	3138.7	3276.3	3456.5
	s				6.2872	6.5390	6.7801	6.9878	7.2338

From Keenan, J. H., Keyes, F. G., Hill, P. G., and Moore, J. G., *Steam Tables — Metric Units,* John Wiley & Sons, New York, 1969. Copyright © 1969, John Wiley & Sons. Reprinted by permission of John Wiley & Sons, Inc.

INDEX

A

Absorption
  of fat, II: 108
  of water, II: 49, 105, 106
Acetals, II: 3
Acids, II: 3, 96
  amino, see Amino acids
  citric, II: 86
  fatty, II: 5, 64
Activation energy, II: 129
A/D, see Analog to digital
Adams Corp., I: 4
Advantages of extruders, I: 1—3
Aerobic plate count, II: 124
Afflatoxin, II: 115
Agency for International Development (AID), II: 132, 138
AID, see Agency for International Development
Akron Extruders Inc., I: 154
Alarm operators, I: 174
Aldehydes, II: 3
Alginates, II: 65
Alignment of protein molecules, II: 96, 103, 106
Alpha-amylase, II: 51
Alpha-D-glucose, II: 41
Amber durum, II: 19
American Corn Milling Federation, II: 49
Amide bonds, II: 102
Amino acids, II: 100—102, 113, 116—118
  fortification of, II: 113
  patterns of, II: 113, 127
  sequence of, II: 101
Amorphous nature of proteins, II: 101
Amperage measurements, I: 116
Amylographic test, II: 49, 53
Amylopectin, II: 41, 42, 45, 51, 63
Amylose, II: 41, 42, 45, 51, 63
Analog to digital (A/D) converter, I: 173
Analogs, II: 89, 99
Analysis
  of costs, I: 181
  of data, I: 122—123
  of response surface (RS), I: 123—124
  of variance (AOV), I: 122
Anderson-Ibec, I: 129, 131—133
Angle of attack, I: 160
Animal feeds, I: 4, 136, 138, 145
Annular die, I: 144
Antinutritional factors, II: 115—116, 131
Antioxidants, II: 5, 116
AOV, see Analysis of variance
Apparent viscosity, I: 25, 26
Appearance, I: 109, 119
Appropriate Engineering Manufacturing Co., I: 143—144
Aqueous ethanol-soluble carbonate, II: 49
Archer Daniels Midland, I: 4
Archimedes screw, I: 7
Arrhenius-type temperature dependence, II: 129

Artificial flavor, see Synthetic flavor
Aspergillus flavus, II: 115
Assembly, I: 165—116
Assumptions, I: 48
Autogenous extruders, I: 4, 129, 143, 166; II: 125, 139
Autolyzed yeast extracts (AYE), II: 4
Automatic control, I: 136
Auxillary extrusion equipment, I: 159—162
Average residence teim, I: 98
Axial area of screw channel, I: 13, 14
AYE, see Autolyzed yeast extracts

B

Baby foods, II: 114
Back pressure, I: 57; II: 25
Bacon-flavored snack, II: 99
Bacteria, II: 108, 124
Baked goods, see also Dried goods, II: 126
  collets, II: 68, 71, 73
Baker-Perkins Inc., I: 156
Balling, II: 23
Barley, II: 61
Barrels, I: 14—16, 132, 149, 152, 155, 166, 175; II: 25, 28—29, 70, 71
  assembly of, I: 165
  design of, I: 108
  extruder, I: 14—16
  grooved, I: 16, 58—59, 68, 127, 138, 152, 178; II: 70, 71
  jacketed, I: 132
  life of, I: 179
  liner of, I: 15, 140, 179
  long, I: 128
  segmented, I: 15, 136
  temperature of, I: 66, 117
  wear on, I: 178—179; II: 29
Base, II: 96
Batch feed system, I: 10
Beans, II: 65, 130, 131
Bearings, I: 178, 179
  inspection of, I: 177
Beaterbars, I: 144
Beta-amylase, II: 42
Beta-carotene, II: 12
Beverages, II: 114, 126, 134, 140
  bases for, I: 138
BHA, see Butylated hydroxyanisol
BHT, see Butylated hydroxytoluene
Binding free gossypol, II: 126
Bins, I: 9
Biopolymers, I: 27, II: 101
Birefringence, II: 43
Blades, II: 34
Blended foods, I: 4, 143; II: 10, 114, 126—145
  indigenous, II: 134
Blenders, I: 158

F

Glanded cottonseed, II: 114, 126
Glandless cottonseed, II: 114
Glandless cottonseed flour, II: 108
Globular protein, II: 102
Glucoamylase, II: 42
Glucose, II: 41
Gluten, I: 147; II: 20, 61, 63
Glycogen, II: 41
Glycosides, II: 131
    linkages of, II: 42
Goitrogenic activity, II: 115
Gossypol, II: 108, 114, 116, 126, 141
Gains, I: 156, 171
Granulation, II: 61
GRAS, see Generally Regarded As Safe
Gravimetric feeders, I: 111, 112; II: 21
Grease seals, I: 178
Griffith Laboratories, II: 99
Grinders, II: 131
Grinding, II: 71, 119, 124, 140
Grits
    cereal, II: 52
    corn, I: 68, 76, 128; II: 52—55, 68, 69, 70, 135
    sorghum, II: 55, 56
    soy, I: 69; II: 95
Grooves, I: 16, 68, 127, 138, 152, 178; II: 70, 71
    angle of, I: 58—59
Gruel, II: 140
Guillotine cutters, I: 134, 160
Gums, II: 77

Hemagglutinates, II: 93
Herschel-Buckley model, I: 23
High-moisture doughs, II: 76—78
High-moisture extrusion, I: 129
High-pressure forming extruders, I: 127, 128
High protein foods, II: 113—145
High-shear cooking extruder, I: 128
High temperature/short time (HTST), I: 1, 3, 128; II: 114, 138
History
    of food extrusion, I: 3—5
    of RTE cereals, II: 79
Holding tank, II: 72
Hollow screws, I: 14, 132, 146
Hollow tubes, I: 162
Hoppers, I: 9
HTST, see High temperature/short time
HVP, see Hydrolyzed vegetable protein
Hydration, II; 96
    value of, II: 105
Hydraulic pullers, I: 170
Hydraulic radius, I: 84
Hydrocolloids, I: 65, 66
    with potato doughs, II: 78
Hydrogen, II: 102
Hydrogen bonds, II: 43, 102
Hydrolyzed vegetable protein, (HVP), II: 4
Hydrophilic ends, II: 64
Hydrophobic ends, II: 64
Hydrostatic pressure, II: 12

H

Hagen-Poiseuille equation, I: 34, 54
Half-products, II: 74—75
Hammer mills, I: 135; II: 134
Hardened material of construction, I: 15
Hartig, I: 68
Head pressures, I: 169
Heat, I: 7, 78, 189; II: 151
    capacity for, see Heat capacity
    dry, II: 120
    flux of, I: 78
    of fusion, I: 79, 81
    of gelatinization, II: 47
    transfer of, see Heat transfer
Heat capacity, I: 189; II: 151
    of water, I: 194; II: 156
Heated press, II: 91
Heaters, I: 16
Heating screw, I: 66
Heat transfer, I: 87, 167, 181, 190
    coefficient of, I: 58; II: 152
Heat treatment, I: 132
    HTST, II: 114
    measurement of, II: 93—94
Height to diameter (H/D) ratio, I: 14
Helical path of food particles, I: 93
Helix, II: 41
    angle, of, I: 13, 80; II: 25
    structure of, II: 41, 45

I

Impact, I: 119
INCAPARINA, II: 138
Incompressible doughts, I: 48, 49
Independent variables, I: 107—108
Indigenous blended food, II: 134
Individual control loops, I: 174
Inert material, II: 84
Infants, II: 132
    food for, see Baby food
Inferential control, I: 171
Inferential meters, I: 112
Ingredients, I: 6, 48, 107, 181 II: 19—21, 61—66, 69
    availability of, II: 136
    cost of, II: 136
    feeders of, II: 21—23
    of plant protein, II: 91—95
    variations in, I: 169
Injection
    of dye, I: 100, 116
    of steam, I: 7, 131, 167, 176, 181
    of water, I: 131, 176
Input of power, I: 72
    Input torque, I: 74
    Inspection, I: 177—178
    Instantaneous pulse, I: 96
Insta-Pro, I: 145; II: 119, 140
Instron, I: 35

Milton Keynes UK
Ingram Content Group UK Ltd.
UKHW051952071024
449327UK00026B/2273